Using Aeronautical Charts

Using Aeronautical Charts

Terry T. Lankford

McGraw-Hill

New York Chicago San Francisco Lisbon London Madrid
Mexico City Milan New Delhi San Juan Seoul
Singapore Sydney Toronto

The McGraw-Hill Companies

Library of Congress Cataloging-in-Publication Data

Lankford, Terry T.
Using aeronautical charts / by Terry T. Lankford.
p. cm.
Includes index.
ISBN 0-07-139117-7
1. Aeronautical charts. I. Title.
TL587.L36 2002
629.132'54—dc21 2002035974

1 2 3 4 5 6 7 8 9 DOC/DOC 0 9 8 7 6 5 4 3 2

ISBN 0-07-139117-7

The sponsoring editor for this book was Shelley Ingram Carr, the editing supervisor was Daina Penikas, and the production supervisor was Pamela A. Pelton. It was set in the Gen1AV1 design in Garamond by Wayne Palmer of McGraw-Hill Professional's Hightstown, N.J. composition unit.

Printed and bound by RR Donnelley.

This book is printed on recycled, acid-free paper containing a minimum of 50% recycled, de-inked fiber.

Contents

Before we take off

My goal, like that of the Federal Aviation Administration (FAA) and the aviation industry as a whole, is the safe and efficient use of our aircraft within the National Airspace System (NAS). Safety is, to a large extent, based on a pilot's training and experience. Unfortunately, the definition of "learning by experience" is "where the test comes before the lesson." This book is dedicated to one aspect of a pilot's training—the application of aeronautical charts and publications to our flying activity.

Before we begin we would like to challenge the reader with a series of true/false and multiple choice questions based on the interpretation and application of aeronautical charts and publications.

True/false

1. 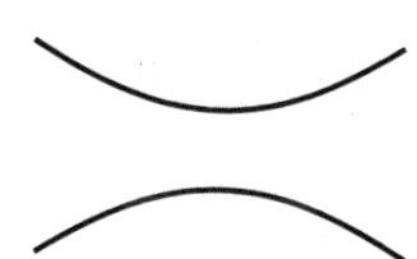

   The mountain pass symbol is always oriented with the direction of the pass.
2. An underlined frequency indicates the facility operates on a part-time basis.
3. The symbol 122.1R indicates the ATC facility has only receive capability on this frequency. Therefore, the pilot must transmit on 122.1 MHz and listen on another frequency—usually the associated navigational aid.
4. Controlled airspace establishes visual flight rules (VFR) minimums, and may mandate minimum pilot qualifications and aircraft equipment.
5. 

   This symbol indicates a visual checkpoint that is stored in the Global Positioning System (GPS) database.

6. VFR charts are often more complex than those used for instrument flight rules (IFR) navigation.
7. Pilots operating under IFR are required to accept charted visual flight procedures (CVFP) when assigned by Air Traffic Control.
8. All information concerning an instrument approach is contained in the published instrument approach procedure (IAP) chart.
9. Operating runway lights are required for night instrument approach operations.
10. Pilots can expect to receive all relevant information required by regulations, during a briefing from a flight service station (FSS) or direct user access terminal (DUAT).
11. The Aeronautical Information Manual (AIM) provides the pilot with mandatory procedures.

Multiple choice

1. Which statement best describes a swamp?
   a. An area of land and water, with abundant vegetation.
   b. A lake with trees growing out of it.
   c. A wooded tract of land that rises above an adjacent marsh.
2. A maximum elevation figure is defined as:
   a. The highest elevation in any group of related relief formations.
   b. The highest elevation, including terrain or other vertical obstacles, bounded by a grid on a chart.
   c. A point on the chart where elevation is noted, usually the highest point on a ridge or mountain range.
3. On average, how many changes occur with each sectional aeronautical chart revision?
   a. 95
   b. 125
   c. 280
4. Terminal area charts (TACs) are designed:
   a. For visual navigation of slow- to medium-speed aircraft.
   b. To allow a pilot to safely navigate in the vicinity of, and remain clear of, congested airspace.
   c. For visual navigation by moderate-speed aircraft and aircraft operating at higher altitudes.

5. Sectional charts are designed:
   a. For visual navigation of a low- to medium-speed aircraft.
   b. To allow a pilot to safely navigate in the vicinity of, and remain clear of, congested airspace.
   c. For visual navigation by moderate-speed aircraft.
6. Minimum climb rate on instrument charts is based on the aircraft's:
   a. True airspeed
   b. Indicated airspeed
   c. Ground speed
7. Which of the following minimum categories applies to most general aviation GPS equipped aircraft?
   a. LNAV/VNAV
   b. LNAV
   c. Circle to land
8. Which of the following are recent improvements to National Aeronautical Charting Organization (NACO) charts?
   a. Briefing strip
   b. Terrain
   c. Graphic missed approach
9. An often-overlooked element of instrument approach procedures are the:
   a. Missed approach procedures
   b. Circling minimums
   c. Notes
10. Pilots must be familiar with which of the following prior to beginning an instrument approach?
    a. The type and use of the equipment on board the aircraft.
    b. All aspects of the approach procedure.
    c. Aircraft approach configuration and missed approach procedure.
11. The official definition of night is:
    a. Sunset to sunrise.
    b. The time between the end of evening civil twilight and the beginning of morning civil twilight.
    c. One hour after sunset to one hour before sunrise.

As you proceed through the book, keep these questions in mind. They will be answered and explained—you may be surprised.

Answers

True/false

1. F
2. F
3. T
4. T
5. T
6. T
7. F
8. F
9. T
10. F
11. F

Multiple Choice

1. b
2. b
3. c
4. b
5. a
6. c
7. b, c
8. a, b, c
9. c
10. a, b, c
11. b

Introduction

An essential part of flight preparation is the acquisition, interpretation, and application of aeronautical information. A primary source of these data are contained in aeronautical charts and their related publications.

Any map user seeks information; a pilot needs specific details on terrain, airspace, landing area, and navigational aids. The cartographer's task is to design a map with the least distortion for the intended purpose. The success of a map is dependent upon the cartographer's and user's knowledge and on their joint realization of the purpose and limitations of the map.

No matter how short or simple the flight, regulations place the responsibility for flight preparation on the pilot. To effectively use available resources during preflight planning, a pilot must understand what information is available, how it is distributed, and how it can be obtained and applied. During the flight, a pilot must constantly use charted information for navigation and communications.

Almost at the beginning of commercial aviation it became apparent that charts could not contain all of the vast and varied information necessary for a safe and efficient flight. Because of the expense of chart production, charts could not economically be updated at every change. To this end, aeronautical publications were developed. Publications primarily provide planning information, data to update charted information, or data that remain relatively unchanged.

In this text, technical concepts and terms are explained in nontechnical language, progressing beyond decoding and translating, to interpreting and applying information to actual flight situations. Discussions include applying chart information to visual flight rules (VFR) and instrument flight rules (IFR) operations, plus low-and high-level flights.

Chart limitations are discussed. There is an examination of chart information and publication usage while flight planning using direct user access terminals (DUATs). This is a sound foundation for the novice and a practical review for the experienced pilot; a thorough knowledge of aeronautical charts and their relationship to the air traffic control system is essential to a safe, efficient operation.

The strictly VFR pilot should not overlook the information in the chapters on instrument charts. The VFR pilot might want to utilize products designed for instrument flying. For example, some pilots supplement visual charts with enroute low-altitude charts for the enroute chart information (airway radials, minimum enroute altitudes, ATC frequencies). Instrument approach procedure charts feature all related communication and navigation frequencies with an airport sketch. A detailed diagram is available for larger airports, which can be very valuable to the VFR pilot. With the increased number of runway incursions, every pilot should obtain a copy of the airport sketch or airport diagram, especially when planning a flight into an unfamiliar or major airport.

The importance of chapters explaining aeronautical publications cannot be overstated. Publications are chart extensions, as important as the charts themselves. Supplemental publications provide information on how to use the air traffic control system, airspace, and procedures. Various commercial supplements are available that go beyond aeronautical data. Many provide commercial information about airport services, transportation, and lodging.

This book provides information that is required to pass written and practical examinations, prepare the dispatcher as well as the pilot—from student through airline transport employee, flying aircraft from recreational airplane to business jet—to operate safer and more efficiently within an increasingly complex environment.

Chapter 1 begins with the limitations of the cartographer. How do we transfer a globe onto a flat surface, then locate a specific point? The various chart projections, each with its own advantages and limitations, are discussed, along with limitations common to all charts; methods and criteria used by the cartographer are explained. The problems of scale, simplification, and classifications are considered. Another section discusses the problem of currency. In an ever-changing environment, how does a pilot obtain the latest information required for a safe and efficient flight? A pilot must be fully aware of the system used to update aeronautical information. This includes the information available from flight service stations (FSSs) and DUAT weather briefings, as well as the services available through commercial chart producers, and their limitations.

Chapters 2 through 4 pertain to visual charts. A pilot has access to charts covering the world. These charts are of little value without the knowledge and understanding to apply their vast wealth of information. These chapters begin with the terminology and symbols used on visual charts, then go on to explore visual charts used in routine flying, and supplemental charts available for other parts of the world and those used for special purposes. Dozens of different charts have been developed for aviation. Each has its own criteria, purpose, and limitations; each is analyzed. The pilot can then apply the array of charts available, with respect to regulations, type of operation, and pilot ability, to efficiently operate within today's aviation system.

Visual chart terminology and symbols are discussed. Topographical features of terrain, hydrography (water and drainage features), culture (man-made objects), and obstruction are defined and chart symbology explained. Next, aeronautical terminology and symbols for airports, navigational aids (NAVAIDS), and airspace are considered. From this analysis, and a practical application of terminology and symbols, the reader should be able to define and explain all visual chart symbols, and apply their meaning to various flight situations.

A thorough analysis of standard visual charts includes visual planning and sectional charts, terminal area charts (TACs), and world aeronautical charts (WACs), which are used most often for VFR flying. A pilot should be able to choose the best product for the planned flight; inappropriate chart selection has often led to pilot disorientation, and, unfortunately, at times to fatal accidents.

Civil aeronautical charts for the United States, its territories, and possessions are produced by the National Aeronautical Charting Office (NACO), which is part of the Federal Aviation

Administration's (FAA) Office of Aviation Systems Standards (AVN). The National Imagery and Mapping Agency (NIMA) provides charts for the rest of the world. Both NACO and NIMA charts use similar symbology and format. NIMA products are of limited value in the United States because of the NACO chart series; however, some NIMA charts might be helpful for planning, or other specialized missions. Special NACO charts are also presented, such as the Grand Canyon VFR aeronautical chart, which provides guidance through canyon airspace affected by a special regulation. Charts and related material published by foreign countries and private publishers are explored. This should give the reader a sense of the products available, but sometimes overlooked.

Chapters 5 through 7 examine instrument procedure charts. Since the late 1960s, more and more pilots are obtaining an instrument rating. General aviation aircraft are approaching and exceeding commercial aircraft performance of the 1940s and 1950s. Each chart series is analyzed with respect to use, regulations, and limitations.

This book is not intended to make recommendations regarding chart publishers, merely to provide the reader with information for an educated choice. It is the pilot's responsibility to select the chart and information publisher that best suits his or her needs.

A number of points can be inferred from these chapters. The FAA and other government agencies have gone to great lengths to establish a safe airway system. But the pilot's safety, and that of the passengers and those on the ground, ultimately rests with a pilot's chart knowledge. The pilot must know what equipment is available and operational on the aircraft. Certain routes and procedures can be flown only with specific equipment, such as automatic direction finder (ADF), distance measurement equipment (DME), or the Global Positioning System (GPS). The pilot must then correctly interpret the chart, refusing any ATC clearance that cannot be complied with because of equipment limitations or deficiencies.

The instrument chart chapters contain interrelated material; therefore, the reader is occasionally asked to refer to a previous chapter, usually a specific figure number. I apologize for any inconvenience, but this seemed the best way to integrate the material.

The IFR-specific chapters begin with enroute instrument charts. This might seem like starting in the middle, with subsequent chapters on departure and arrival procedures, and approaches; however, because most terminology and symbols on enroute charts also apply to the other chart series it would seem to be a logical starting point. A solid understanding of enroute products is a foundation for examining specific charts used for departure, arrival, and approach.

Departure procedures (DP) and standard terminal arrival route (STAR) charts are discussed. Once the domain of air carrier, corporate, and military pilots, and contained in separate publications, DPs and STARs are an integral part of today's IFR system, and published along with their associated instrument approach procedures. A pilot's acceptance of a DP or STAR is an agreement to comply with the requirements of the procedure. In addition to terminology and symbols, procedural requirements are explored. For example, what is the pilot expected to do when radio communications are lost during a radar vector DP or STAR?

Instrument approach procedure (IAP) charts are dissected. Each item of information is decoded, defined, and explained. A half-dozen different instrument approach procedures, including RNAV and GPS approaches, are analyzed with respect to information available, pilot procedures and requirements, and lost communications. A thorough understanding of approach procedures should provide the pilot with the knowledge to apply any approach to a flight situation.

The final chapters discuss and analyze publications that support aeronautical charts. Recall that aviation complexities forced selected information off the chart and into supplemental publications. Aeronautical publications that support NACO VFR and IFR charts are discussed. These publications, which are direct extensions of charts, provide the detailed information beyond the scope of charts, and serve as interim documents to update charts between publication cycles or during periods of temporary, but extended, outages. Details on how to obtain NACO aeronautical charts and related publications are provided.

Additional supplement information available to the pilot is discussed. These documents should not be overlooked. They include Department of Defense (NIMA) flight information publications, Canadian supplements, additional documents available through NACO, and those produced by private vendors.

Appendix A contains a decode list of Notice to Airmen (NOTAM) and chart contractions. Appendix B provides a list of agencies that produce and distribute aeronautical charts and publications. Included are addresses, telephone numbers, Internet sites, and e-mail addresses for supplement product providers, many of which offer free catalogs.

The following chapters, hopefully told with a little humor and practical examples of application, explain how to use, translate, interpret, and apply aeronautical charts. Incidents and anecdotes are not intended to disparage or malign any individual, group, or organization; the sole purpose is illustration.

Navigation principles and techniques are not within the scope of this book; therefore, they are not included. The reader should remember that, especially in aviation, the only thing that doesn't change is change itself. Everything possible has been done to ensure that the information in this text is current and accurate at the time of writing.

Using Aeronautical Charts

1

Projections and limitations

The map or chart is a unique means of storing and communicating topographic and other information. It allows the user to reduce the world to a size within the normal range of vision. It is the most effective means of relating the relative location, size, direction, and distance of locations and objects on the earth.

For our purposes we will define *map* as a graphic representation of the physical features—natural, artificial, or both—of the earth's surface, by means of signs and symbols, at an established scale, on a specified projection, and with a means of orientation. A chart is a special-purpose map designed for navigation; specifically an aeronautical chart is a chart designed to meet the requirements of aerial navigation.

Ptolemy, a second century A.D. Egyptian mathematician, astronomer, and geographer, developed principles for map making, including divisions and coordinates. Ptolemy's reference system used lines parallel with and at equal intervals from the equator to the poles, and lines north and south at right angles to the parallels, equally spaced along the equator: latitude and longitude. Ptolemy acknowledged his work was based on Hipparchus (second century B.C.) who was the first geographer to use parallels of latitude and meridians of longitude. Interestingly, Columbus, more than 1000 years later, based his reasoning that if you sailed west you would reach the east on Ptolemy's assumption that the world was round.

In the sixteenth century, Gerardus Mercator began working on the problem of transferring a sphere, the earth, onto a flat map. He reintroduced latitude and longitude, which had only been sporadically used. His maps allowed the navigator to draw a straight line between locations and determine a constant course. In 1569, he introduced his first map of the world using the Mercator projection.

By the eighteenth century, the German mathematician, astronomer, and physicist Johann Heinrich Lambert improved the cone projection derived from Ptolemy to conform with two standard parallels. He developed a half-dozen projections, one of which was Lambert conformal conic; however, it was more than a century before cartographers fully appreciated the value of these projections, and not until World War I was it adopted by the allies for military maps.

Fact

With commerce growing, it became apparent that map makers needed a common reference for coordinates. To this end, in October 1884 the first International Meridian Conference adopted the Observatory of Greenwich, England, as the prime meridian. Longitude was reckoned in two directions, east and west of the Greenwich meridian.

The original aeronautical chart service used the Lambert conformal projection while the Navy's Hydrographic Office employed the Mercator projection. For planning and operating within a global system, new projections were introduced and old systems adapted. The Lambert conformal was preferred for its accuracy in air navigation for most parts of the world; however, for navigation in polar regions, charts using the transverse Mercator or polar stereographic projections were selected.

Today's aerial navigators have a variety of highly accurate aeronautical charts that cover the world. Aeronautical charts and chart making have made tremendous progress since their inception at the beginning of the twentieth century. But, like aviation weather and weather products, aeronautical charts have limitations that must be understood by the pilot. A sound understanding of chart purpose and limitations are necessary to safely and efficiently apply the assortment of charts in an operational environment.

Only a globe can accurately portray locations, directions, and distances of the earth's surface. The earth is not actually a sphere, but a spheroid. The earth only approximates a true sphere because of the force of rotation that expands the earth at the equator and flattens it at the poles. This elliptical nature of the earth is a concern to cartographers; for most practical purposes of navigation, the earth can be considered a sphere. On the earth, meridians of latitude are straight and meet at the poles; parallels of longitude are straight and parallel, as shown in Fig. 1-1. Meridian spacing is widest at the equator and zero at the poles; parallels are equally spaced. Scale is true for every location.

Because a globe is not possible for practical aeronautical charts, mathematicians and cartographers have devised a number of systems, known as projections, to describe features on the earth in the form of a plane or flat surface; several have already been mentioned. A map projection is

1-1 *Only on a globe are areas, distances, directions, and shapes true.*

a system used to portray the sphere of the earth, or a part thereof, on a plane or flat surface.

Locations on the earth are described by a system of latitude and longitude coordinates. By convention, latitude is named first, then longitude. Refer to Fig. 1-1. The reference point for latitude is the equator, with latitude measured in degrees north and south of the equator. Longitude is measured east and west of the prime, or Greenwich, meridian.

Any point on the earth can be described using the system of latitude and longitude in degrees, minutes, and seconds. (There are 60 minutes per degree and 60 seconds per minute.) With the sophistication of today's navigational systems, aeronautical charts and publications can express latitude and longitude in degrees, minutes, and hundredths of a minute. A degree is an arc 1/360 of a circle; therefore, a point with latitude 47°N longitude 122°W would be the intersection of the parallel 47° north of the equator and the meridian 122° west of the Greenwich meridian. Degrees can be further subdivided into minutes (′), which represent 1/60 of a degree, and seconds (″), which represent 1/60 of a minute. For example, the Seattle-Tacoma International Airport is located N47°26.28′ W122°18.67′ (47 degrees 26.28 minutes north; 122 degrees 18.67 minutes west).

Speculation

Many have wondered why a circle is divided into 360 degrees, a degree into 60 minutes, and a minute into 60 seconds? Why not divide a circle into 100 units or 400 units? It appears the reason for the selection was due to the mathematical sophistication at the time when this system was originally developed. Three hundred and sixty can be divided evenly by the most whole numbers (i.e., 2, 3, 4, 5, 6, 8, 9, 10, etc.).

Each degree of latitude equals 60 nautical miles (nm). Because meridians meet at the poles, a degree of longitude decreases in length with distance north or south of the equator; therefore, only for the special case of the equator does a degree of latitude equal 60 nautical miles.

Projections

The goal of the map projection is to accurately portray true areas, shapes, distances, and directions. This includes the condition that lines of latitude are parallel and meridians of longitude pass through the earth's poles and intersect all parallels at right angles.

- **Areas.** Any area on the earth's surface should be represented by the same area at the scale of the map. These projections are termed *equal area* or *equivalent.*
- **Distances.** The distance between two points on the earth should be correctly represented on the map. These projections are termed *equidistant.*
- **Directions.** The direction or azimuth, from any point to any other points on the earth should be correct on the map. These projections are termed *azimuthal* or *zenithal.*
- **Shapes.** The shape of any feature should be correctly represented. The scale around any point must be uniform. These projections are termed *conformal.*

Because it is possible to obtain all these properties only on a globe, the cartographer must select the projection that preserves the most desired properties on the basis of the chart's use. Figures 1-2 through 1-4 illustrate the three projections most often used on aeronautical charts: Mercator, Lambert conic conformal, and polar stereographic. Table 1-1, Chart Projections, compares the different projections used on aeronautical charts.

Mercator projection

The Mercator projection transfers the surface of the earth onto a cylinder tangent at the earth's equator. This is illustrated in Fig. 1-2. On the Mercator projection, meridians and parallels appear as lines crossing at right angles. Meridian graticule spacing, the network of parallels and meridians forming the map projection, is equal and parallel as shown in Table 1-1 and Fig. 1-2.

Meridians are parallel on the Mercator projection, unlike meridians on the earth that meet at the poles. This results in

Table 1-1. Chart projections.

Projection	Lines of longitude (meridians)	Lines of latitude (parallels)	Graticule spacing	Scale	Uses
Globe/earth	Straight and meet at poles	Straight and parallel	Meridian spacing maximum at equator, zero at poles; parallels equally spaced	True	Impractical for navigation
Mercator	Straight and parallel	Straight and parallel	Meridian spacing equal; parallel spacing increases away from the equator	True only along equator; distortion increase away from equator	Dead reckoning; celestial
Lambert conic conformal	Straight con-verging at poles	Concave arcs	Parallels equally spaced	True along standard parallels	Pilotage; dead reckoning
Polar stereographic	Straight radiating from the poles	Concentric circles; unequally spaced	Conformal	Increases away from pole	Polar navigation

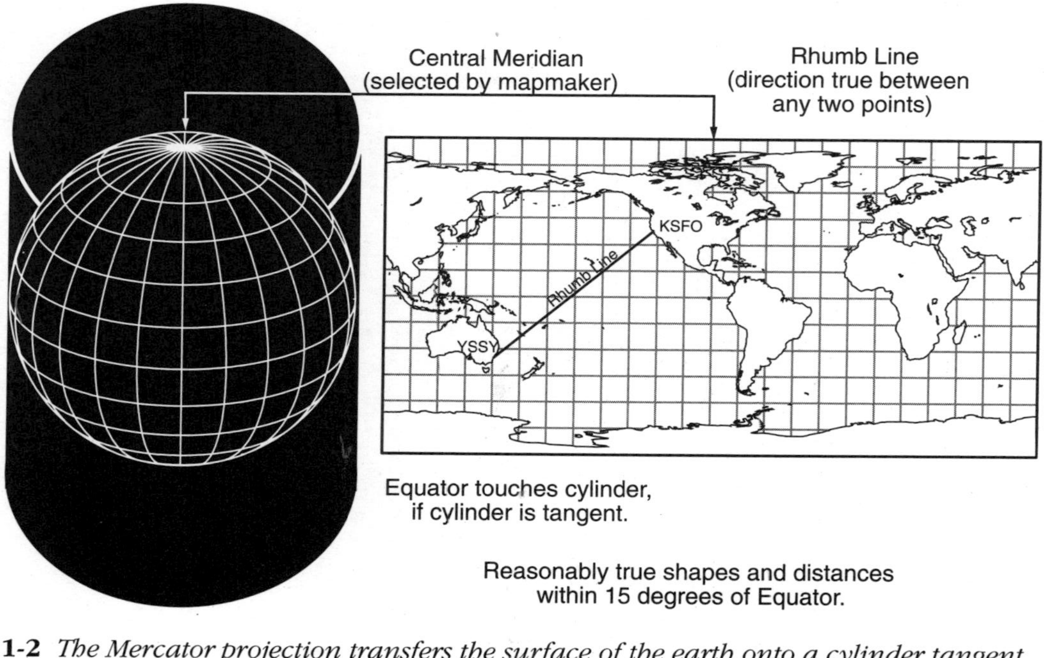

1-2 *The Mercator projection transfers the surface of the earth onto a cylinder tangent at the earth's equator; meridians and parallels appear as lines crossing at right angles.*

increasingly exaggerated areas toward the poles. Scale, which is the relationship between distance on a chart and actual distance, changes with latitude.

The advantage of the Mercator is that a straight line on this projection crosses all meridians at the same angle. This allows the navigator to set a constant course from one point to another. A course crossing all meridians at a constant angle is known as a rhumb line. Figure 1-1 shows a rhumb line from San Francisco, United States (KSFO) to London, England (EGLL). Figure 1-2 illustrates a rhumb line from Sydney, Australia (YSSY) to San Francisco (KSFO).

A rhumb line is not normally the shortest distance between two places on the surface of the earth. The great circle distance is always the shortest distance between points on the earth. A great circle is an arc projected from the center of the earth through any two points on the surface. A great circle arc is illustrated in Fig. 1-1. A great circle, unlike the rhumb line, crosses meridians at different angles, except in two special cases: where the two points lie along the equator or the same meridian. In both of these cases the rhumb line and great circle coincide.

For practical purposes at low latitudes, rhumb and great circle distances are nearly identical. As latitude and distance increase, differences become increasingly significant, as illustrated in Fig. 1-1. However, the projection is conformal in that angles and shapes within any small area are essentially true.

Lambert conformal conic projection

A projection widely used for aeronautical charts is the Lambert conformal conic, with two standard parallels. This is illustrated in Fig. 1-3. As the name implies, a cone is placed over the earth and intersects the earth's surface at two parallels of latitude. Scale is exact everywhere along the standard parallels; between the parallels scale decreases and beyond the parallels scale increases. Distortion of shapes and areas are minimal at the standard parallels, but distortion increases away from the standard parallels.

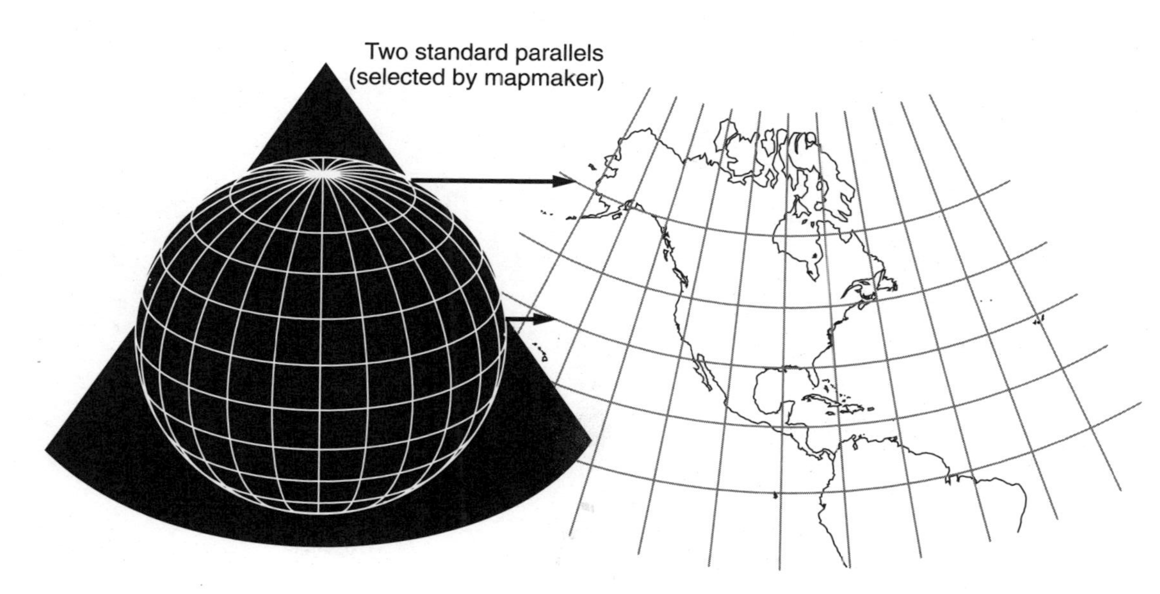

1-3 *With the Lambert conformal conic, a cone is placed over the earth and intersects the earth's surface at two parallels of latitude.*

All meridians are straight lines that meet at a point beyond the map; parallels are concentric circles. Meridians and parallels intersect at right angles. The chart is considered conformal because scale is almost uniform around any point; scale error on any chart is so small that distances can be considered constant anywhere on the chart. A straight line from one point to another very closely approximates a great circle.

Polar stereographic projection

The standard Lambert is too inaccurate for navigation above a latitude of approximately 75° to 80° north or south of the equator. The polar stereographic projection is sometimes used for polar regions. A plane tangent to the earth at the pole provides the projection. This is illustrated in Fig. 1-4. Meridians are straight lines radiating from the pole and parallels are concentric circles. A rhumb line is curved and a great circle route is approximated by a straight line. Directions are true only from the center point of the projection. Scale increases away from the center point. The projection is conformal, with area and shape distortion increasing away from the pole.

Horizontal datum

Cartographers need a defined reference point upon which to base the position of locations on a chart. This is known as the horizontal datum, or horizontal constant datum, or horizontal geodetic datum. The horizontal datum used as a reference for position is defined by the latitude and longitude of this initial point. Prior to 1992 the horizontal datum for the United States was located at Meades Ranch, Kansas, referred to as the North American Datum 1927 (NAD 27).

With the introduction of geodetic satellites for mapping the earth's surface and satellite navigation systems for innumerable applications, there were recommendations to revise NAD 27. Beginning October 15, 1992, the horizontal geodetic referencing system used in all charts and chart products was changed from the North American Datum of 1927 to the North American Datum of 1983 (NAD83). This resulted in differences of only approximately 1000 ft between NAD 27

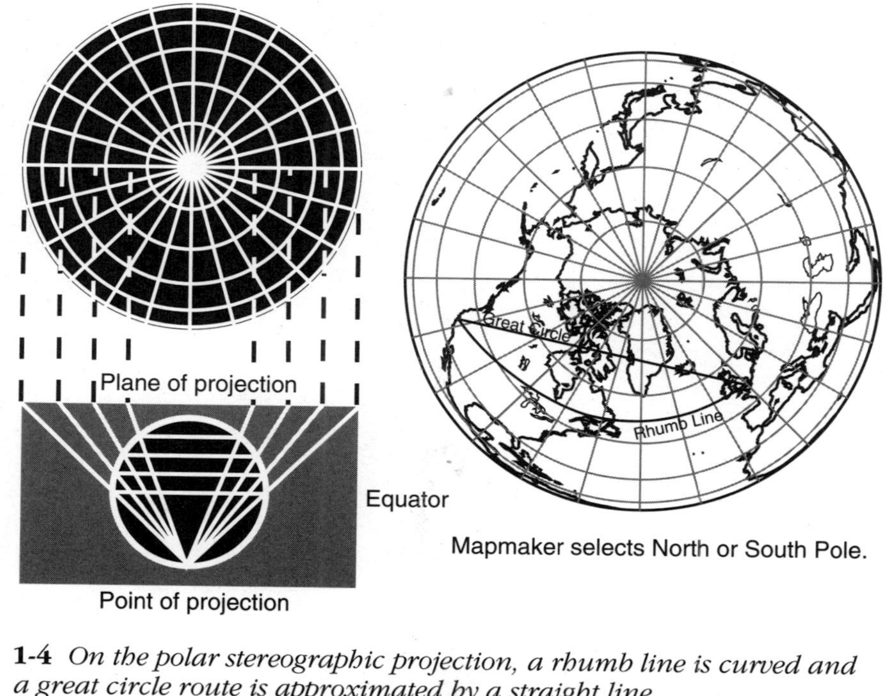

1-4 *On the polar stereographic projection, a rhumb line is curved and a great circle route is approximated by a straight line.*

and NAD 83 positions. The greatest coordinate shifts occurred in Alaska and Hawaii where latitude was moved by as much as 1200 ft and longitude by up to 950 ft. In the conterminous United States, the maximum change was approximately 165 ft in latitude and 345 ft in longitude.

Limitations

In addition to the limitations of chart projections, cartographers and chart users are faced with the problems of scale, simplification, and classification. Finally, chart users, especially pilots, are faced with the tremendous issue of current information. With visual charts updated only annually or semiannually, and instrument charts every 28 to 56 days, the pilot must understand the system used to provide the latest information between routine chart revisions.

Scale

Charts provide a reduced representation of the earth's surface. Scale defines the relationship between a distance on a chart and the corresponding distance on the earth. Scale is generally expressed as a ratio: the numerator, customarily 1, represents chart distance, and the denominator, a large number, represents horizontal ground distance. For example: 1:500,000 (sometimes written 1/500,000) states that any unit, whether inch, foot, yard, statute mile, nautical mile, or kilometer, on the chart, represents 500,000 units on the ground. That is, one inch on the chart equals 500,000 inches on the earth.

Chart makers provide scales for conversion of chart distance to statute or nautical miles or kilometers. Manufacturers of aeronautical plotters provide scales for standard aeronautical charts; however, the pilot must be familiar with the plotter used and chart scale for accurate calculations.

Case Study

While flying in the annual Hayward–Bakersfield–Las Vegas air race we use sectional charts (1:500,000), except in the Las Vegas area where a terminal area chart (1:250,000) is available. Sure enough, in my haste I measured a leg using the wrong scale. This is dis-

astrous in a race where the finishing order is a matter of seconds and tenths of a gallon of fuel.

Oh yes! When the FAA designs written test questions requiring the pilot to measure chart distance on a sectional chart (1:500,000), one of the answers is always the distance measured using the world aeronautical chart (1:1,000,000) scale on the plotter.

The smaller the scale of a chart, the less detail it can portray. For example, a chart with a scale of 1:1,000,000 cannot provide the detail of a chart with a scale of 1:250,000. Charts with a smaller scale increase the size of the area covered (assuming a constant size), but reduce the detail that can be depicted.

Simplification and classification

Because scale reduces the size of the earth, information must be generalized. Making the best use of available space is a major problem in chart development. The detail of the real world cannot be shown on the chart. The crowding of lines and symbols beyond a specific limit renders the chart unreadable, yet the amount of information that might be useful or desirable is almost unlimited. The smaller the chart scale, the more critical and difficult the problem; therefore, the cartographer is forced to simplify and classify information.

Simplification is the omission of detail that would clutter the map and prevent the pilot from obtaining needed information. The necessity for detail is subjective and not all will agree on what should, or should not, be included. The inclusion of too much detail runs the risk of confusing the reader by obscuring more important information. For example, the chart producer might have to decide whether to include a prominent landmark, the limit of controlled airspace, or a symbol indicating a parachute jump area. The problem of simplification has led directly to the use of aeronautical publications, such as the *Airport/Facility Directory.*

Figure 1-5 illustrates simplification. In Fig. 1-5 picture A depicts a world aeronautical chart with a scale of 1:1,000,000; picture B shows a sectional chart with a scale of 1:500,000; picture C illustrates a terminal area chart (TAC) with a scale of 1:250,000. The inset boxes in A and B cover the same lateral area as the TAC in picture C. Note the additional detail available with the larger scale. The disadvantage is that a relatively large chart covers a relatively small geographical area. Therefore, as we shall see, various charts are designed for specific uses, each with it own purpose of limitations.

The *Airport/Facility Directory* is divided into seven booklets that cover the United States, including Puerto Rico and the Virgin Islands. Alaska is covered by the *Alaska Supplement,* areas of the Pacific by the *Pacific Chart Supplement.* These directories are a pilot's manual containing data on airports, seaplane bases, heliports, navigational aids, communications, special notices, and operational procedures. They provide information that cannot be readily depicted on charts such as airport hours of operation, types of fuel available, runway widths, lighting information, and other data, as well as a means to update charts between editions.

Classification is necessary in order to reduce the amount of information into a usable form. The cartographer must classify towns, rivers, and highways of different appearance on the ground into a common symbol for the chart. The pilot must then be able to interpret this information.

Case Study

On a flight from Phillipsburg, Pennsylvania, to Huntington, West Virginia, we were forced to fly below a 1500-foot overcast because of the weather and radio trouble. Because of poor planning on my part, we had only a world aeronautical chart, instead of the larger-scale sectional chart. I misidentified the Kanawha River as the Ohio River. To verify the position I checked the highways, railroads, and power lines adjacent to the river. Nothing matched. After a few moments of utter confusion I checked the time from last known position and determined we could not have made it all the way to the Ohio. Based on this estimate of distance, I reevaluated our position as over the Kanawha; now everything matched. On the WAC the Kanawha was

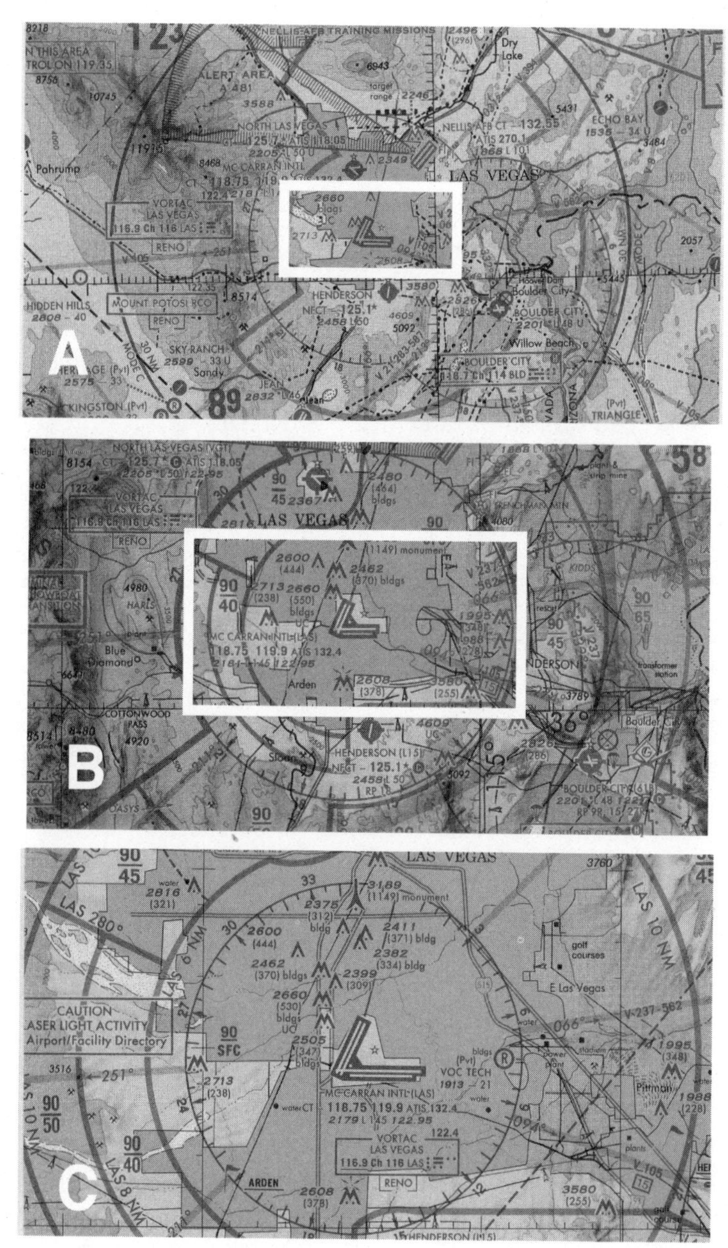

1-5 *The inset boxes in A and B cover the same lateral area as the TAC illustrated in C; note the additional detail available with the larger scale.*

represented by a thin blue line, the Ohio by a wide blue line. From the air both rivers appeared identical.

To maximize the amount of information on a chart the cartographer uses symbols. Symbol shape, size, color, and pattern are used to convey specific information. The pilot must be able to interpret these symbols. Lack of chart symbol knowledge can lead to misinterpretation, confusion, and wandering into airspace where one has no business—and has done so.

Figure 1-5 illustrates the depiction of Class B airspace on the three chart scales. Picture A, the WAC, can depict only the outer lateral limits of the Las Vegas Class B airspace. Picture B, the sectional, contains the Class B areas with their bases and tops. Picture C, the TAC, provides much more topographical detail, as well as the arc distances of the various concentric areas.

Currency

With an ever-changing aeronautical environment, charts are outdated almost as soon as they are printed and become available. A pilot's first task when using any chart or publication is to ensure its currency.

Visual charts are revised and reissued semiannually or annually. Changes to visual charts are supplemented by the "Aeronautical Chart Bulletin," in the *Airport/Facility Directory,* revised every 56 days.

National Aeronautical Charting Office (NACO) IFR charts contained in the Terminal Procedures Publication (TPP) are published every 56 days. A 28-day midcycle change notice (CN) volume contains revised procedures that occur between the 56-day publication cycle. (A list of airports in the CN is available on the NACO Web site at; www.naco.faa.gov.) These changes are in the form of new charts. The subsequent publication of the TPP incorporates change notice volume revisions and any new changes since change notice issuance.

With the increased use of aeronautical databases for both VFR and IFR navigation, it's extremely important that pilots understand how their equipment is updated. Aeronautical databases are typically updated on a 28-day cycle. (Normally the new database becomes effective at 0901Z on the date of the change.) The pilot's database must not be updated and flown prior to the effective data and time, nor the equipment used for navigation without the update after the update's effective date and time.

Like databases, some chart providers issue charts to their subscribers prior to their effective date. Pilots must be careful not to attempt to use charts prior to their effective date.

Visual and instrument charts are further supplemented by the FAA's Notice to Airmen system (NOTAMs). The NOTAM publication is published every 28 days and supplements the "Aeronautical Chart Bulletin" in the *Airport/Facility Directory,* The TPP, and change notice volumes. A detailed discussion of the NOTAM publication is provided in Chap. 8.

Aeronautical information not received in time for publication is distributed through the FAA's telecommunications systems. These include unanticipated or temporary changes, or hazards when their duration is for a short period or until published. A NOTAM is classified into one of three groups:

- NOTAM (D)
- NOTAM (L)
- FDC NOTAM

NOTAM (D)s consist of information that requires wide distribution. NOTAM (D)s contain information that might influence a pilot's decision to make a flight, or require alternate routes, approaches, or airports. They are considered "need-to-know" and issued for certain landing area restrictions, lighting aids, special data, and air navigation aids that are part of the National Airspace System (NAS). NOTAM (D)s are issued for all public-use airports listed in the *Airport/Facility Directory.*

NOTAM (L)s include information that requires local dissemination, but does not qualify as a NOTAM (D). The criteria for NOTAM (L)s have changed significantly over the last decade. NOTAM (L)s have always been a bastard operation. However, with the introduction of DUATs most previously NOTAM (L) criteria items now receive NOTAM (D) distribution, for example, all public use airports and most tower light outages. Tower light outages that do not meet the criteria for NOTAM (D) are disseminated as a NOTAM (L). That is any obstruction 200 ft AGL or less and more than 5 statute miles from a public-use airport.

FDC NOTAMs consist of information that is regulatory in nature pertaining to charts, procedures, and airspace. They include conditions that fall into the following categories: interim IFR flight procedures, temporary flight restrictions, 14 CFR Part 139 certificated airport condition changes—airport rescue and fire fighting (ARFF) required for air carrier operations, emergency flight rules special data, and laser light activity.

NOTAMs from each category are routinely provided as part of a standard flight service station (FSS) weather briefing. NOTAMs, except (L), are also available through direct user access terminals (DUATs) and other commercial weather vendors. When FDC NOTAMs are associated with a specific facility identifier, they are included as part of the DUAT briefing; however, most enroute chart changes are not associated with a specific facility identifier. DUAT users are faced on every briefing with a disclaimer: "FDC NOTAMs that are not associated with an affected facility identifier will now be presented unless you specifically choose to decline to receive such information." It's almost like looking for the proverbial needle in a haystack, but to be safe, these NOTAMs must be reviewed. DUAT users must remember that, once FDC NOTAMs are published, unlike the FSS briefings, where the pilot has the option to request the data, "Published FDC NOTAM Data are not available, and must be obtained from other publications/charts/etc." Pilots using NACO charts must be aware of these limitations and plan their flight briefings accordingly. This might mean a call to the FSS specifically for any pertinent FDC NOTAMs.

Once a new chart becomes effective, the NOTAMs, including those carried in the *Notice to Airmen* publication and "Aeronautical Chart Bulletin" of the *Airport/Facility Directory,* are canceled and removed. Pilots are presumed to be using current charts.

A major advantage of commercial suppliers of chart services is a more timely revision schedule than is available for most government products. Immediate and short-term changes to the National Airspace System must still be obtained, but this information is normally provided as part of an FSS standard briefing or DUAT briefing.

Case Study

A British friend with whom I flew in England in the middle 1960s was astonished by the frequent revisions to United States aeronautical charts. The British civil charts we had to fly with were infrequently updated and only contained about a fifth of the information of United States charts. For example, all NAVAIDs were shown by a single symbol without any indication of the type or frequency. And, oh yes, during this period the English charts cost about $5 in U.S. currency—at the time U.S. charts went for about 35¢ each! I'm afraid those days are gone forever.

NACO has made every effort to ensure that each piece of information shown on its charts and publications is accurate. Information is verified to the maximum extent possible. According to NACO, "You, the pilot, are perhaps our most valuable source of information. You are encouraged to notify NACO of any discrepancies you observe while using our charts and related publications. Postage-paid chart correction cards are available at authorized chart sales agents for this purpose (or you may write directly to NACO, at the address below). Should delineation of data be required, mark and clearly explain the discrepancy on a current chart (a replacement copy will be returned to you promptly)."

FAA, National Aeronautical Charting Office
AVN-510, SSMC4, Sta. #2335
1305 East West Hwy.
Silver Spring, MD 20910-3281
9-AMC-Aerochart@faa.gov
1-800-626-3677

NACO emphasizes that the "Use of obsolete charts or publications for navigation may be dangerous. Aeronautical information changes rapidly, and it is vitally important that pilots check the effective dates on each aeronautical chart and publication to be used. Obsolete charts and publications should be discarded and replaced by current editions."

NACO cites the following reasons why pilots need to fly with current charts.

- Each sectional chart averages 280 changes every 6 months.
- Each *Airport/Facility Directory* averages 780 changes every 56 days.
- Each Terminal Procedures Publication volume averages 75 changes every 56 days.

Case Study

A pilot called the FSS and requested a briefing from Bishop to Santa Cruz, California. The briefer explained that the airport was closed. The pilot responded, "Oh, I must be using an old chart." Indeed, the airport had been closed for over 2 years. There are no valid reasons for using obsolete charts.

The use of current charts and publications, and the need for obtaining a complete or standard preflight briefing, cannot be overemphasized. It's like using the rest room before a flight, we know we should, but sometimes it's a little inconvenient. Failure, however, has often led to a very uncomfortable flight. Only by understanding the system can pilots ensure they meet their obligation as pilot in command.

2

Visual chart terminology and symbols

Terms and symbols discussed in this chapter not only apply to standard visual charts—world aeronautical charts, sectional charts, and terminal area charts—but also most carry over to planning charts, instrument charts, and charts produced by the National Imagery and Mapping Agency (NIMA) and other chart producers. Throughout this chapter the reader should keep in mind chart purpose and limitations, and the problems facing the cartographer, specifically: scale, simplification, and classification as discussed in Chap. 1.

Topography and obstructions

The elevation and configuration of the earth's surface are of prime importance to visual navigation. Cartographers devote a great deal of attention to portray relief and obstructions in a clear, concise manner.

Topography is the configuration of the surface of the earth; it is subdivided into hypsography, hydrography, and culture. Hypsography is the science or art of describing elevations of land surfaces with reference to a datum, usually sea level; it is that part of topography dealing with relief, or elevation of terrain. Hydrography is the science that deals with the measurements and description of the physical features of the oceans, seas, lakes, rivers, and their adjoining coastal areas, with particular reference to their use for navigational purposes. It is that part of topography pertaining to water and drainage features, where drainage features are those associated with shorelines, rivers, lakes, marshes, and the like, and the overall appearance made by these features on a map or chart. Culture is features of the terrain that have been constructed by man: roads, buildings, canals, and boundary lines.

Relief is defined as inequalities of elevation and the configuration of land features on the surface of the earth. Relief might be represented on charts by contours, colored tints representing elevations or gradients (hypsometric tints), shading, spot elevations, or short disconnected lines drawn in the direction of the slopes (hachures).

Obstructions are man-made vertical features that might affect navigable airspace. The National Aeronautical Charting Office (NACO) maintains a file of more than 50,000 obstacles in the United States, Canada, Mexico, and the Caribbean. Normally, only obstacles more than 200 ft above ground level (AGL) are charted, unless the obstacle is considered significant, for example, near an airport or much higher than surrounding terrain.

Terrain

Three methods are used on aeronautical charts to display relief: contour lines, shaded relief, and color tints. Contour lines, as the name implies, connect points of equal elevation above mean sea level (MSL) on the earth's surface. Contours graphically depict terrain and are the principal means used to show the shape and elevation of the surface. Contours are depicted by continuous lines—except where elevations are approximate, then with broken lines—labeled in feet MSL. On sectional charts basic contours are spaced at 500-ft intervals, although intermediate contours might be shown at 250-ft intervals in moderately level or gently rolling areas. Occasionally, auxiliary contours portray smaller relief features at 50-, 100-, 125-, or 150-ft intervals.

Figure 2-1 shows how contours, shaded relief, and color tints depict terrain. Contours show the direction of the slope, gradient, and elevation. For example, in Fig. 2-1 the valley floors have little or no gradient, while the mountains have steep gradients. The contours are labeled with their elevation. A good example is Lite Peak located just to the northwest of the "Contours" label in Fig. 2-1. Note that the valley has an elevation of just over 7000 ft. The contours surrounding Lite Peak range from 8000 ft, just above the valley floor, to a maximum elevation of 10,093 ft. Shaded relief depicts how terrain might appear from the air. The cartographer shades the areas that would appear in shadow if illuminated from the northwest. Shaded relief enhances and supplements contours by drawing attention to canyons and mountain ridges. Shaded relief is illustrated in Fig. 2-1. Color tints depict bands of elevation. These colors range from light green for the lowest elevations to dark brown for higher elevations. Color tints in Figure 2-1 range from light brown in the valley to dark brown—you'll have to take my word for it—over the mountain ranges, supplementing the contours and enhancing recognition of rapidly rising terrain.

In addition to contours, shading, and tints, significant elevations are depicted as spot elevations, critical elevations, and maximum elevation figures (MEFs). Spot elevations represent a point on the chart where elevation is noted. Figure 2-1 shows an example of spot elevation. Terrain rises gradually from the plateau to the southeast toward the shaded relief of the mountain crests. In this example the spot elevation is 8240 ft. Spot elevations may indicate the highest point on a ridge or mountain range, or, as in the example, the highest elevation within an area of gradually rising terrain. A solid dot depicts the exact location when known. A × denotes approximate elevations; where elevation is known, but location approximate, only the elevation appears, without the dot or × symbol. Critical elevation is the highest elevation in any group of related and more-or-less similar relief formations. Critical elevations are depicted by larger elevation numerals and dots than are used for spot elevations. Figure 2-1 illustrates the difference between spot and critical elevations. An example of a critical elevation in Fig. 2-1 is the highest elevation along the mountain crest at 14,047 ft. (We certainly wouldn't want to attempt to cross this ridge in our Turbo-Cessna 152—that's an attempt at a little humor—or a Piper Warrior.)

Maximum elevation figures represent the highest elevation, including terrain and other vertical obstacles—natural and man-made—bounded by the ticked lines of the latitude/longitude grid on the chart. MEFs are intended for quick reference or emergency use. They are depicted to the

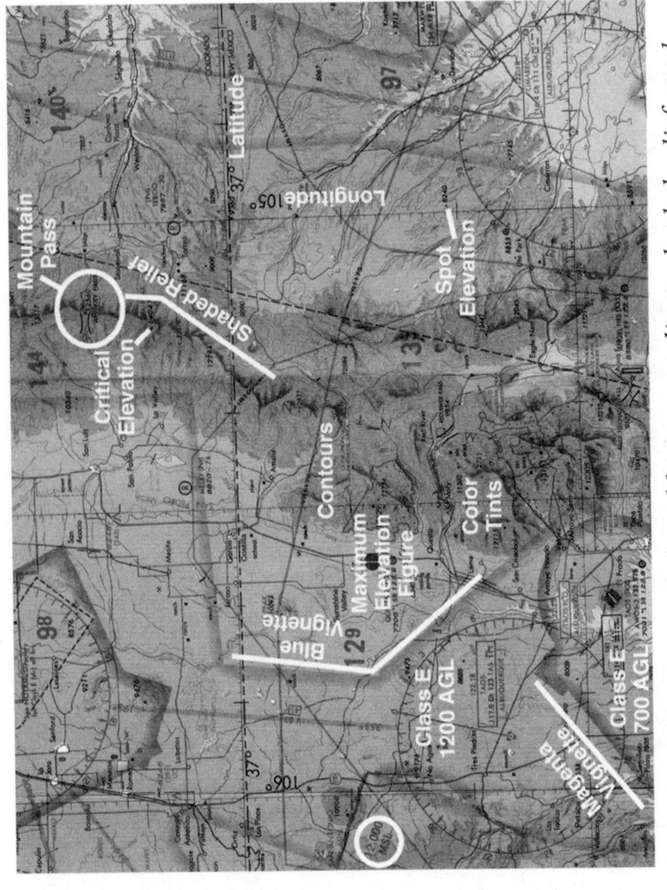

2-1 *Chart relief is represented by contour lines, shaded relief, and color tints.*

nearest 100-ft value; the last two digits of the number are omitted. The center of the grid in the left portion of Fig. 2-1 shows 12^9. (The grid in the illustration is bounded by a latitude of 37° north at the top and a longitude of 106° west of the left. The right and bottom boundaries are the subsequent latitude/longitude boundaries depicted on the chart to the right and below.) The MEF for this grid is 12,900 ft above mean sea level (MSL). This figure is determined from the highest elevation or obstacle, corrected upward for any possible vertical error (including the addition of 200 ft AGL, or 300 ft AGL over yellow city tint, for any natural or man-made obstacle not portrayed), then rounded upward to the next higher hundred-foot level; therefore, almost all MEFs will be higher than any elevation or obstacle portrayed within the grid on the chart. (The highest elevation for the example in Fig. 2-1 is a critical elevation of 12,529 ft.) Pilots should note that MEFs cannot take into account altimeter errors and should be considered like any other terrain elevation figure for flight planning purposes.

Latitude and longitude are labeled in degrees. Lines of latitude and longitude are subdivided by lines representing 10 minutes and half lines representing 1 minute of arc. Because longitude represents the same distance anywhere on the earth, unlike latitude which decreases toward the poles, 1 minute of longitude anywhere on the earth equals 1 nautical mile (nm); therefore, lines of longitude can be used for quick estimates of distance. (Latitude and longitude lines are illustrated in Fig. 2-1. Recall the discussion of MEFs.)

Fact?

After the large earthquakes in southern California in 1971 and 1994 the Los Angeles sectional chart contained the comment: "CAUTION—Terrain elevations subject to change without notice." Well, not really, but maybe they should have.

Notwithstanding the previous observation, certain parts of the country with a relatively flat, rolling landscape contain areas of sharply rising terrain. Such areas may lull the unsuspecting pilot to an unpleasant surprise. To help pilots avoid such areas some charts contain a clear warning: "Caution: Rapidly rising terrain." One such example is contained on the St. Louis sectional chart west of Nashville, Tennessee. The bluffs and cliffs use the symbology shown in Fig. 2-2.

Other topographical relief features considered suitable for navigation are contained in Fig. 2-2. They include sand and gravel areas, rock strata and quarries, mines, craters, lava flows, and other relief information usable for visual checkpoints and navigation. With the exception of sand, which is depicted in magenta (pink), relief features are shown in black.

Feature	Symbol	Feature	Symbol
SAND OR GRAVEL AREAS		ESCARPMENTS, BLUFFS, CLIFFS, DEPRESSIONS, ETC.	
SAND RIDGES (To Scale)		LEVEES AND ESKERS	levee
SAND DUNES (To Scale)		HACHURING	
ROCK STRATA OUTCROP	rock strata	DEPRESSION (Includes Mound within Depression)	2000 1000
QUARRIES (To Scale)	quarry	DISTORTED SURFACE AREAS	lava
STRIP MINE (To Scale)	strip mine	LAVA FLOWS	
MINE DUMPS AND TRAILINGS (To Scale)	mine dump	UNSURVEYED AREAS (Label approx. As Required)	UNSURVEYED
CRATERS	crater	MOUNTAIN PASSES	VPZZZ 12632

2-2 *Supplemental relief features aid in visual flying.*

Lava flows, and sand ridges and dunes, are quite pronounced when seen from the air and make excellent checkpoints, especially if they are isolated by other terrain features. An example of a lava flow is contained in Fig. 2-3. Unfortunately, most of these features only appear in the western United States. Strip mines and large quarries also make excellent checkpoints because of their visibility. A classic example is the Bingham Cooper Mine south-southwest of Salt Lake City, Utah.

2-3 *Lava flows can be distinct geographical features; however, most are found in the western United States.*

Large craters, where they appear, make excellent checkpoints. This is illustrated in Fig. 2-4, northern Arizona's meteor crater. Because of its size, on a flat plateau it is easily recognizable.

Case Study

I was taking a Cessna 182 from California to Wichita, where three other pilots and I would pick up four new Cessna 150s. The Cessna 182 had no navigational radios. East of Prescott, Arizona, we climbed over an undercast and proceeded eastward. My dead reckoning navigation plan was to find a hole or return to Prescott—Prescott FSS had direction finding (DF) equipment. Forty-five minutes into the leg we found a hole and descended. As luck would have it we descended over Arizona's meteor crater, which was just about the most prominent landmark in that area. From there we were able to proceed on course.

Hachures show the direction and degree of slope in hills and other elevations. Occasionally a pilot might see the term

2-4 *Large craters, where they appear, make excellent checkpoints; because of its size and location on a flat plateau, northern Arizona's meteor crater is easily recognizable.*

karst. Karst topography is an area that consists of sinkholes and fissures resulting from underground streams flowing through porous limestone, eroding the rock from below the ground. It indicates an area of distorted terrain, relative to the surrounding geology, that is identifiable from the air. Such an area is located on the WAC chart southeast of Albuquerque, New Mexico.

One other topographical feature should be mentioned. Mountain passes are depicted by black curved lines. The name of the pass and elevation are included. In Fig. 2-1 the Whiskey Pass is shown, upper right, with an elevation of 12,560 ft MSL. It's important for pilots to understand that the pass symbol does not necessarily represent the actual direction of the pass. The highest point on the road surface, determined from United States Geodetic Survey (USGS) data, is used for pass elevation.

Beginning in 2002 GPS receiver databases, mountain pass waypoints are expected to appear on sectional charts. GPS waypoints will indicate the end points of mountain pass routes for VFR purposes. Chart depiction is expected to be similar to that shown in Fig. 2-2, Mountain Passes, "VPZZZ."

Hydrography

Hydrography pertains to water and drainage features. Hydrographic features on aeronautical charts are represented in blue: streams, rivers, and aqueducts are depicted by a single blue line; lakes and reservoirs by a blue tint. To indicate open water, the tint is a dark blue; inland water and waterways are shown with a lighter blue tint. Small dots or hatching indicate where streams and lakes fan out (or are not perennial), or where reservoirs are under construction.

Figure 2-5 shows how shorelines, lakes, streams, reservoirs, aqueducts, and other hydrographical features are depicted. Shorelines usually make excellent checkpoints, except where they are relatively straight without features. Pilots need to pay attention to shoreline orientation. For example, most people assume that California's coastline is north-south; however, in certain areas, such as around

SHORELINES		AQUEDUCTS (Suspended or Elevated) KANATS (With Air Vents)	underground aqueduct
LAKES	Non-perennial 756 Perennial	FLUMES PENSTOCKS & SIMILAR FEATURES	flume flume underground flume
RESERVOIRS & DAMS	Man-made Under construction	FALLS	falls
STREAMS	Perennial Non-perennial	RAPIDS	rapids
STREAMS		CANALS	ERIE abandoned
STREAMS		SMALL CANALS (Perennial) (Non-Perennial)	
SAND DEPOSITS		SMALL CANALS (Abandoned or Ancient) (Numerous)	
AQUEDUCTS	aqueduct abandoned aqueduct underground aqueduct	SALT PANS AND EVAPORATORS (Man Exploited)	salt pans

2-5 *Shorelines, lakes, streams, reservoirs, and aqueducts have landmark value, with specific limitations.*

Santa Barbara, the coast is actually east-west. This has led to much confusion, especially for student pilots and others unfamiliar with the area. Lakes usually make good checkpoints, especially when their shape is unique or they are dammed, as discussed in Chap. 3.

Definite shorelines, those that do not change and have been surveyed, are depicted by a solid outline, as shown in Fig. 2-5. Fluctuating shorelines are shown by a short dashed outline; unsurveyed shorelines are shown with a long dashed outline.

Caution needs to be exercised with all lakes and streams, perennial and nonperennial. A perennial lake or stream contains water year-round; a nonperennial lake is intermittently dry, usually during the dry season. There can be confusion during periods of drought when perennial lakes may dry up. When lakes are too numerous to be depicted individually, they are labeled as "numerous small lakes," shown with representative patterns and described as needed. Nonperennial lakes may encompass areas where the lake remains perennial, as illustrated in Fig. 2-5.

Reservoirs may show a natural shoreline with their associated dams. Man-made shorelines are typically labeled when necessary for clarity.

Streams may be fanned out (alluvial fan), braided, or disappearing. Streams may fluctuate seasonally. When this occurs with undefined limits, the stream is outlined with a dashed line; when seasonal bank limits are prominent and constant, a solid line depicts their maximum extent. This is illustrated in Fig. 2-5.

Discrepancies in depictions result from human decisions to drain, expand, or abandon lakes, reservoirs, and streams. Streams should, if at all possible, be used only to support other checkpoints. That is, there should be other landmarks that establish position, supported by the position of the stream. Perennial and nonperennial streams should be treated with the same cautions as perennial and nonperennial lakes. Reservoirs are similar to lakes, and can be treated in the same way; however, reservoirs are usually perennial.

Flumes, penstocks, and similar features are hydrographic features used to carry water as a source of power, to devices such as waterwheels. Canals are shown to scale when this can be accommodated on the chart. Canal names may be displayed.

Figure 2-6 shows additional hydrographical features used on aeronautical charts. Among these are symbols for swamps, bogs, and marshes; glaciers; and ice and snow fields. Ice peaks, polar ice, and pack ice are features restricted to polar and arctic regions. Boulders, wrecks, reefs, and underwater

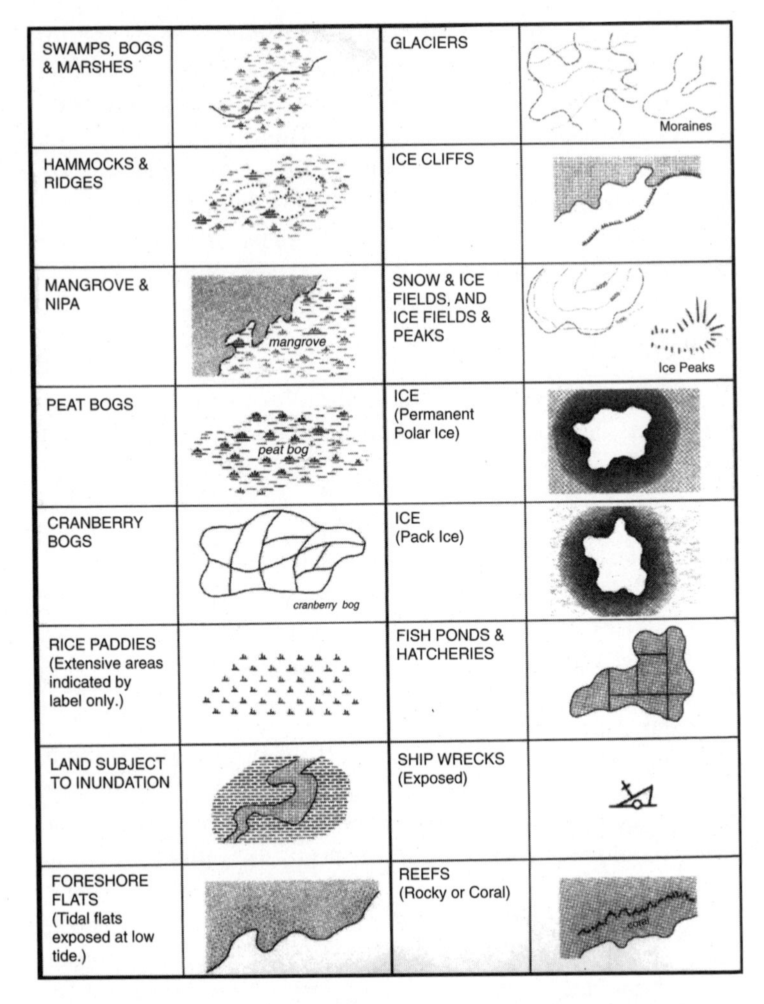

2-6 *Some hydrographical features warn of danger, such as swamps; others provide supplemental landmark value.*

features are displayed because they have certain landmark value. Some of these features might be small and difficult to identify.

Case Study

The terrain for a flight from Lufkin, Texas, to New Orleans' Lakefront Airport was flat, but contained widespread areas of swamp. I normally plan direct flights, unless there is a significant reason to do other-

wise. Having spent most of my life in the West, I didn't know what a swamp was. Fortunately, a concerned specialist at Lufkin FSS (now decommissioned) explained the significance of this hydrographical feature. A swamp is nothing more than a lake with trees growing out of it. He was certainly right and I appreciated the information. Pilots flying over unfamiliar terrain, as in this case, would be well advised to seek the advice of local pilots or the local FSS—as local as possible during these times.

Hummocks and ridges describe a wooded tract of land that rises above an adjacent marsh or swamp. Mangroves are any of a number of evergreen shrubs and trees growing in marshy and coastal tropical areas; a nipa is a palm tree indigenous to these areas. Bogs are areas of moist, soggy ground, usually over deposits of peat. Tundra describes a rolling, treeless, often marshy plain, usually associated with arctic regions. (Tundra does not have a symbol; areas of tundra are indicated by the word *tundra* on the chart.)

Miscellaneous underwater features, not otherwise symbolized, are enclosed by a dashed line. The type of feature is labeled; for example, "shoals." Other than canals, the other features in Fig. 2-6 might be difficult to verify and should normally only be used to support other checkpoints.

Pilots flying in northwestern Montana and especially Alaska can expect to see glaciers and glacial moraines (debris carried by the glacier), ice cliffs, snow and ice fields, and ice caps. Figure 2-7 contains two photographs of glacier activity in Alaska. The top photo is a glacier in the vicinity of Juneau. The lower photo shows an ice cliff from a glacier in Glacier Bay.

Culture

Land features constructed by man include populated areas, roads and highways, railroads, buildings, canals, dams, boundary lines, and the like. Many landmarks that can be easily recognized from the air, such as stadiums, racetracks, pumping stations, and refineries, are identified by brief descriptions adjacent to a small black square or circle marking

2-7 *Most ice features are applicable only to arctic regions.*

their exact location. Depictions might be exaggerated for improved legibility. Figure 2-8 contains a description of cultural topographic information on aeronautical charts. Differences between sectionals and WACs are noted.

Populated areas—built-up urban areas—are outlined in accordance with their coverage on the ground as seen from the air,

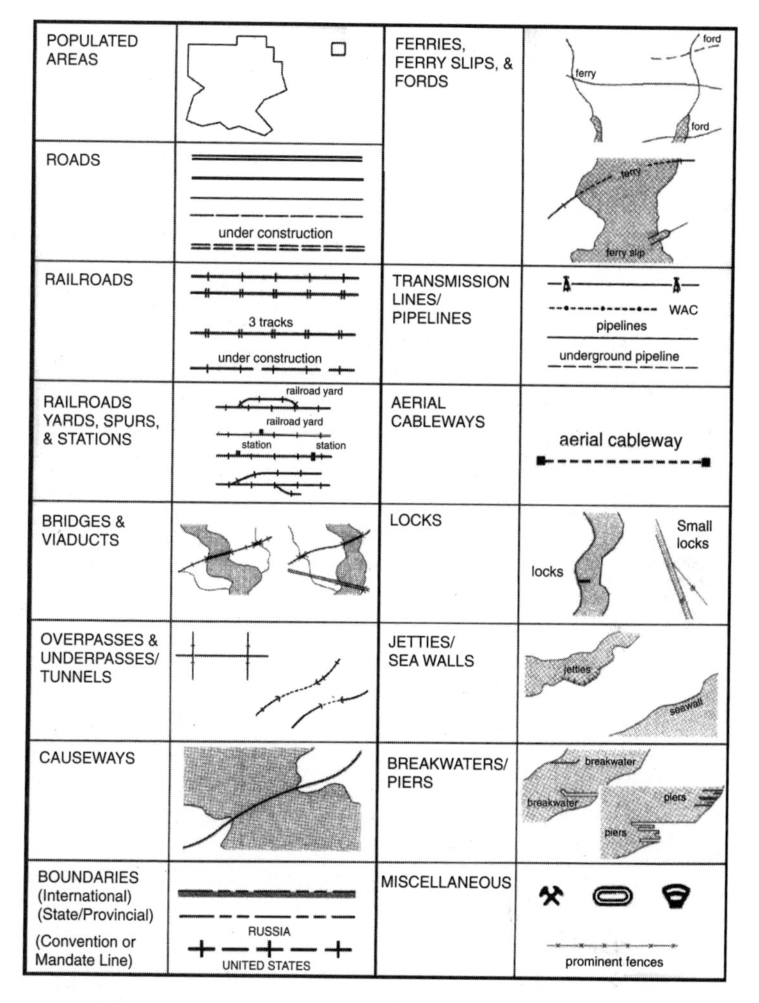

2-8 *Major cities, roads, railroads, and bridges make good visual landmarks.*

in a yellow tint. Large cities are termed category 1. Cities and large towns are described as category 2. Category 2 populated areas are indicated by a rectangle as illustrated in Fig. 2-8, depicted on the chart with a yellow tint. The name of a category 1 or 2 populated areas is shown. Towns and villages are classified as category 3. These areas are indicated by a small black circle or square on aeronautical charts.

Like populated areas, roads are shown according to their size or category, depicted in black. Refer to Fig. 2-8. Category 1 dual-lane highways are shown as double back line. Category 2 primary roads are depicted with a single, thick black line. Category 2 secondary roads are represented by a single thin black line. Category 3 trails are shown as a single, dashed black line. (Note that a single, dashed line may also represent a dismantled railroad when combined with the label "dismantled railroad.") Category 1 and 2 roads may also be identified by their name (Lincoln Highway), or U.S. route or interstate highway number.

Single-track railroads have one crosshatch, double and multiple railroads have a double crosshatch. When more than two tracks exist, the number of tracks may be labeled, as shown in Fig. 2-8. Electric railroads are labeled as such, "electric." Railroads often make excellent checkpoints. A word of caution. Numerous railroads emanate like spokes from many large cities. Pilots navigating exclusively by the "iron compass" have become hopelessly confused when they inadvertently took the wrong track—pardon the pun.

Aerial cableways can be extremely hazardous. Pilots must exercise additional caution when in the vicinity of these hazards, especially at low altitude and in low visibilities.

Miscellaneous cultural symbols consist of shaft mines and quarries (indicated by crossed picks), racetracks, outdoor theaters, and prominent fences. Other landmark areas are outlined with a black dashed line and labeled appropriately.

Pilots should never navigate solely by one landmark. Major highways (category 1) also make excellent checkpoints, but they suffer from the same problems as the railroad. Category 2 roads, and especially secondary category 2, are often difficult to positively identify, especially when flying over sparse areas of desert or plains. Bridges, viaducts, and causeways are often very good checkpoints.

Large and medium cities are shown by their outline as it appears on the ground. This helps significantly with identification. Towns and villages are represented only by a small circle. Especially where several towns or villages are in the same general area, this symbology makes them hard to positively identify. Political boundaries are shown by standard map symbols. Cultural coastal features are depicted because of their landmark value.

Pilots should pay particular attention to the symbol for aerial cableways—formally called catenaries. Aerial cableways, conveyers, etc., depicted on aeronautical charts, are a cable, power line, cable car, or similar structure suspended between peaks, a peak and valley below, or across a canyon or pass. A cableway is normally 200 feet or higher above terrain, which poses a very serious hazard to low-flying aircraft. It may be marked with orange balls or lights.

Cultural features are not revised as often as aeronautical information; therefore, especially in areas of rapid metropolitan development, cultural features as seen from the air may differ from those depicted on the chart.

Power transmission lines (high-tension lines) are depicted for their landmark and safety value. Often, transmission lines can be used to verify the identification of other landmarks. Although not normally qualifying as an obstruction, their depiction alerts pilots flying at low altitudes to this sometimes almost invisible hazard.

When chart features are too small to show their detail, a small circle or square is used to indicate their location. These objects are described and labeled as required. Below is a list of such topographical features:

- Reservoirs
- Numerous canals and ditches
- Tundra
- Springs, wells, and waterholes
- Isolated rocks (bare or awash)
- Fish hatcheries
- Towns and villages
- Time zones
- Stadiums
- Forts
- Cemeteries
- Oil wells
- Water tanks
- Gas tanks
- Lookout towers
- Coast Guard stations

Lookout towers may be identified by their site number (P-17) and include the elevation of the base of the tower (618). Coast Guard stations are labeled with the contraction CG.

Obstructions

Obstruction symbols are used to depict man-made vertical features that may affect the national airspace system. Charted obstructions normally consist of features extending higher than 200 ft AGL. Objects 200 ft or lower are charted only if considered hazardous, for instance, close to an airport where they might affect takeoffs and landings. Except for helicopters, regulations require that aircraft pilots, even over sparsely populated areas, cannot "operate closer than 500 ft to any...structure," except for takeoff or landing; therefore, objects 200 ft high or lower, except in the case of an emergency, should have no significant operational impact. Helicopters may be operated at less than these minimums if the operation is conducted without hazard to persons or property on the surface. Helicopter pilots must understand these charting limitations and their responsibility to operate safely.

Sectional charts contain a caution note: "This chart is primarily designed for VFR navigational purposes and does not purport to indicate the presence of all telephone, telegraph and power transmission lines, terrain or obstacles which may be encountered below reasonable and safe altitudes." The fact that objects at 200 ft can exist without the requirement to be charted should be a sobering thought when you are considering scud running.

Scud Running

Scud is shreds of small detached clouds moving rapidly below a solid deck of higher clouds. Pilots who attempt to negotiate these conditions—scud running—often fly at extremely low altitudes and in low visibility. These operations can be very hazardous to the health of the pilot, passengers, and those on the ground.

Pilots must also keep in mind that many obstructions have guy wires that extend some distance outward from the structure. Helicopter, balloon, and ultralight aircraft pilots need to understand these additional hazards, along with charting limitations, and plan their flights accordingly.

Obstruction symbols are colored blue. Obstacles less than 1000 ft AGL are shown by the standard obstruction symbol as illustrated in Fig. 2-9. Obstacles 1000 ft or higher AGL are shown by the elongated obstruction symbol, as shown in Fig. 2-9. As necessary and for additional landmark value, the type of obstruction may be indicated (building, stack, tank, etc.).

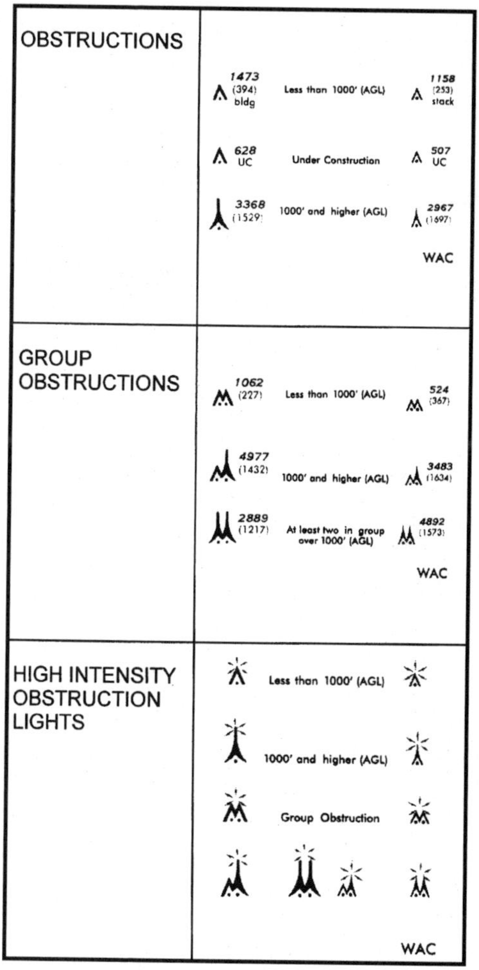

OBSTRUCTIONS	1473 (394) bldg	Less than 1000' (AGL)	1158 (253) stack
	628 UC	Under Construction	507 UC
	3368 (1529)	1000' and higher (AGL)	2967 (1697)
			WAC
GROUP OBSTRUCTIONS	1062 (227)	Less than 1000' (AGL)	524 (367)
	4977 (1432)	1000' and higher (AGL)	3483 (1634)
	2889 (1217)	At least two in group over 1000' (AGL)	4892 (1573)
			WAC
HIGH INTENSITY OBSTRUCTION LIGHTS		Less than 1000' (AGL)	
		1000' and higher (AGL)	
		Group Obstruction	
			WAC

2-9 *Normally, only obstructions higher than 200 ft AGL are charted.*

Height of the obstacle AGL and elevation of the top of the obstacle MSL are shown when known or when they can be reliably determined. The height AGL is shown in parentheses below the MSL elevation of the obstacle—which is a logical arrangement:

2468
(1200)

The height of the obstacle is 1200 ft above ground level (1200). The top of the obstacle above mean sea level is 2468 ft. The height AGL might be omitted in extremely congested areas to avoid confusion. Within high-density groups of obstacles, only the highest obstacle in the area will be shown by the group obstacle symbol. Obstacles under construction are indicated by the letters UC immediately adjacent to the symbol. When available, the eventual AGL height of the obstruction will appear. Obstacles with strobe lighting systems are shown as indicated in Fig. 2-9. It's important to understand that obstruction lights may operate only at night. Additionally, pilots should realize that high-intensity lights may operate part-time.

Aeronautical

Aeronautical information on visual charts consists of airports, radio aids to navigation (NAVAIDs), airspace, and navigational and procedural information. Airports are identified for size (runway lengths), use (civil, military, private), and services. The specific availability of fuel types and grades and repairs is detailed in a directory rather than on a chart. The type and frequency of NAVAIDs are depicted. Controlled- and special-use airspace within the scope of the chart is shown (below 18,000 ft). Navigational information such as magnetic variation, airway intersections, and lighting aids is depicted.

Radio communication and navigation are often depicted as frequencies in megahertz (MHz) or kilohertz (kHz). The conventional ultrahigh-frequency (UHF) transmitters, generally used by the military, and very high frequency (VHF) transmitters—NAVAID band 108.0 MHz to 117.95 MHz—are depicted on aeronautical charts in blue. Low-frequency (LF) and medium-frequency (MF) transmitters—NAVAID band 190 kHz to 535 kHz—are depicted in magenta.

Airports

Visual charts depict civil, military, and some private, landplane, helicopter, and seaplane airports. Airport symbols are depicted in Fig. 2-10. Airports with other than hard-surfaced runways, such as dirt, sod, gravel, and the like, are depicted as open circles. (An open circle is also used to indicate an

LANDPLANE-MILITARY (Refueling and repair facilities for normal traffic.) (All recognizable runways, including some which may be closed, are shown for visual identification.)	WAC	LANDPLANE-CIVIL (Refueling and repair facilities for normal traffic.)	WAC
SEAPLANE-MILITARY (Refueling and repair facilities for normal traffic.)	WAC	SEAPLANE-CIVIL (Refueling and repair facilities for normal traffic.)	WAC
LANDPLANE-CIVIL/MILITARY (Refueling and repair facilities for normal traffic	WAC	SEAPLANE-CIVIL-MILITARY (Refueling and repair facilities for normal traffic.)	WAC
HELIPORT (Selected)	WAC	SEAPLANE-EMERGENCY (No facilities or complete information is not available.)	WAC
LANDPLANE-EMERGENCY (No facilities or complete information is not available.)	PUBLIC USE RESTRICTED OR PRIVATE UNVERIFIED ABANDONED WAC	ULTRALIGHT (Selected)	WAC not shown
		AIRPORT DATA	FSS NO SVFR NAME (NAM) CT - 118.3* AWOS-3 135.425 285 L 72 122.95 VFR Advsy 125.0 FSS NO SVFR NAME CT - 118.3* ATIS 123.8 285 L 72 U Airport of entry WAC

2-10 *Airport symbols provide information on size, use, and services, providing enough data for the pilot to operate into or out of the field under the provisions of VFR.*

airport with a hard-surfaced runway less than 1500 ft in length.) These public-use airports have limited attendance or no services available. Hard-surfaced runways of 1500 to 8069 ft are enclosed within a circle depicting runway orientation. All recognizable runways, including some that might be closed, are shown for visual identification. Hard-surfaced runways greater than 8069 ft do not conveniently fit in a circle; the circle is omitted, but runway orientation is preserved. Airports served by an FAA control tower (CT) or nonfederal control tower (NFCT) are shown in blue, all others in magenta—a purplish red color.

Military airports are depicted in the same general way as civil airports. For other than hard-surfaced runways, military airports are shown with a circle within a circle, as illustrated in Fig. 2-10. Hard-surfaced runways are symbolized in the same way as public-use airports. Military airports are identified by abbreviations such as AFB (Air Force Base), NAS (Naval Air Station), AAF (Army Air Field), and MCAS (Marine Corps Air Station).

Tick marks around the basic airport symbol indicate the availability of fuel and that the airport is attended during normal working hours. Pilots should keep in mind that types of fuel and specific hours attended are contained in the *Airport/Facility Directory,* with changes or nonavailability of services mentioned in NOTAMs.

Airports are plotted in their true geographic position unless the symbol conflicts with a radio navigation aid at the same location. In such cases, the airport symbol will be displaced, but the relationship between the airport and the NAVAID will be retained.

Restricted, private, and abandoned airports are shown for emergency or landmark purposes only. Pilots wishing to use restricted or private landing facilities must obtain permission from that airport's authority. A check of your insurance policy might also be in order. Some policies restrict landings to public-use airports, except in emergencies. Airports are labeled unverified when they are available for public use, but warrant more than ordinary precaution because of lack of current information on field conditions, or available information indicates peculiar operating limitations. Abandoned airports are depicted for their landmark value or to prevent confusion with an adjacent usable landing area. They are normally restricted to paved airports with a least 3000-ft runways.

Selected ultralight flight parks appear only on sectional charts as an F within the airport circle.

Airport data provides the pilot with basic airport information, such as airport name, limited radio-frequency information, and abbreviated airport data. Detailed information is contained in the appropriate *Airport/Facility Directory*. Below is a description of airport data that is illustrated in Fig. 2-10.

☆ A star (☆) associated with the airport symbol indicates a rotating or oscillating beacon located at the aerodrome.

FSS An FAA Flight Service Station (FSS) is located on the field. In the absence of a control tower, or when the tower is closed, the FSS provide Local Airport Advisories (LAAs) for the airport. LAA is a service provided by FSS or the military consisting of information to arriving and departing aircraft concerning wind, favored runway, altimeter setting, known traffic, and airport conditions. This information is advisory and does not constitute an ATC clearance.

RFSS A Remote Flight Service Station provides LAA service for an airport without an FSS on the field. To the extent that information is available, LAA provided by an RFSS is the same as LAA service where the FSS is located on the field. Pilots should not confuse RFSS with an airport that has a Remote Communications Outlet (RCO) served by an FSS. An FSS with an RCO does not provide LAA service. Pilots should use the appropriate Common Traffic Advisory Frequency (CTAF) for airport advisory and traffic information.

NO SVFR NO SVFR informs pilots that fixed wing operations in accordance with special visual flight rules (SVFR) are not authorized within the airport's surface-based controlled airspace.

Ⓡ A circle with the letter R indicates that airport surveillance radar is available. However, the absence of this symbol does not necessarily mean that the tower does not have access to radar data. In fact many towers have access to a radar "bright" display without this symbol in their airport data. This information is not shown on WAC charts.

AIRPORT NAME The airport name is the next item in the airport data. Special airport traffic areas are indicated on the chart by the airport name placed within a box—as in the above example. Two still exist in Alaska at Anchorage and Ketchikan. Specific requirements are contained in the *Alaska Supplement,* Regulatory Notices. With airspace reclassification, Class D airspace has replaced the control zone and airport traffic areas. This has all but eliminated the need for special airport traffic areas defined by 14 CFR Part 93.

(LOCID) Following the airport name is the airport's location identifier (LOCID). This is a group of three or four alphanumeric characters representing the airport's official identification. This is very helpful for pilots with area navigation systems and assisting ATC with the pilot destination for flight following services.

CT-118.3* Control tower (CT) primary frequency. A star (*) indicates part-time operation. Additional tower frequencies and hours of operation are contained in tabulated form on the chart margins and in the *Airport/Facility Directory.*

NFCT NFCT indicates that a nonfederal control tower is in operation at the airport.

Ⓒ A circle with the letter C indicates the airport's Common Traffic Advisory Frequency (CTAF). This information is not shown on WAC charts. This frequency is usually the tower frequency at airports with part-time towers.

ATIS (frequency) The availability of Automatic Terminal Information Service (ATIS) and the specific frequency for this service is listed in the airport data.

AWOS-3 (frequency) AWOS indicates an automated weather observing system is available for the airport. AWOS is only shown when a full-time ATIS is not available. AWOS is not shown on WAC charts.

Automated Observations

AWOS was originally operationally classified into four levels. These levels consisted of AWOS-A, which reports altimeter setting only; AWOS-1 reporting temperature, dew point, wind, and altimeter setting; AWOS-2 adds visibility information; and AWOS-3 reports sky conditions and ceiling along with the other elements.

285 The next element of airport data is field elevations. Field elevation is never abbreviated; it is indicated in feet above sea level. In the example, field elevation is 285 ft MSL.

L (*L) The letter L indicates that airport lighting operates from sunset to sunrise. All lighting codes refer to runway lights. The lighted runway might not be the longest or might not be lighted full length. Specific details on which runways have lights and availability of other airport lighting is contained in the *Airport/Facility Directory*. An asterisk (*) preceding the letter "L" means that lighting limitations exist—operating for limited hours or pilot controlled. Here again, the pilot must check the *Airport/Facility Directory* for details.

72 The next element of the airport data is the length of the longest runway. Runway length is the length of the longest active runway including displaced threshold, and excluding overruns. Runway length is shown to the nearest 100 feet, using 70 as the division point; a runway 8070 ft long is charted as 81, and a runway 8069 ft long is charted as 80. In the example, the longest runway is 7200 ft.

122.95 An aeronautical advisory station or UNICOM frequency may be included. WAC charts will indicate the availability of this service with only the letter U. UNICOM is a nongovernment communication facility. These stations are often operated by airport management or the local fixed base operator (FBO). At uncontrolled airports these frequencies may provide local weather and airport information.

VFR Advsy VFR advisory service is shown where ATIS is not available and the frequency is other than the primary control tower frequency.

RP 5, 30 The last airport data entry contains traffic pattern information. Nonstandard (right) patterns are indicated by the letters RP, followed by the affected runways.

In the example, the airport has right patterns from runways 5 and 30. Left patterns are implied for all other runways by their omission from this data entry. An asterisk (*) refers the pilot to the *Airport/Facility Directory* for details.

Remarks Other remarks are added as required, such as airport of entry. When airport data information is not available, the respective character is replaced by a dash. For example, ...**285** — *72* 122.95....In this case airport lighting information is unreliable or not available.

Radio aids to navigation

Radio aids to navigation consist of any electronic means of navigation. Until recently the most common have been the low-frequency (LF) and medium-frequency (MF) radio beacons, very high frequency omnidirectional radio range (VOR), and distance measuring equipment (DME). Since the mid-1970s, various area navigation systems have been developed, with the global positioning system (GPS) one of the most recent advances.

Area Navigation

Area navigation (RNAV) consists of ground-based or satellite equipment that provides enhanced navigational capability to the pilot. RNAV equipment can compute the airplane's position, actual track and ground speed, and then provide meaningful information relative to a route of flight. Typical equipment will provide the pilot with distance, time, bearing, and crosstrack error related to a selected waypoint. Several distinctly different navigational systems with different navigational performance characteristics are capable of providing area navigational functions. Present day RNAV includes inertial navigation systems (INS), LORAN, VOR/DME, and GPS systems.

Figure 2-11 shows standard symbols for the very high frequency omnidirectional radio range, a VOR collocated with distance measuring equipment (VOR/DME), and a VOR collocated with a tactical air navigation (TACAN) facility. A TACAN provides azimuth information similar to a VOR, but on an ultrahigh frequency (UHF) band used by military aircraft, and

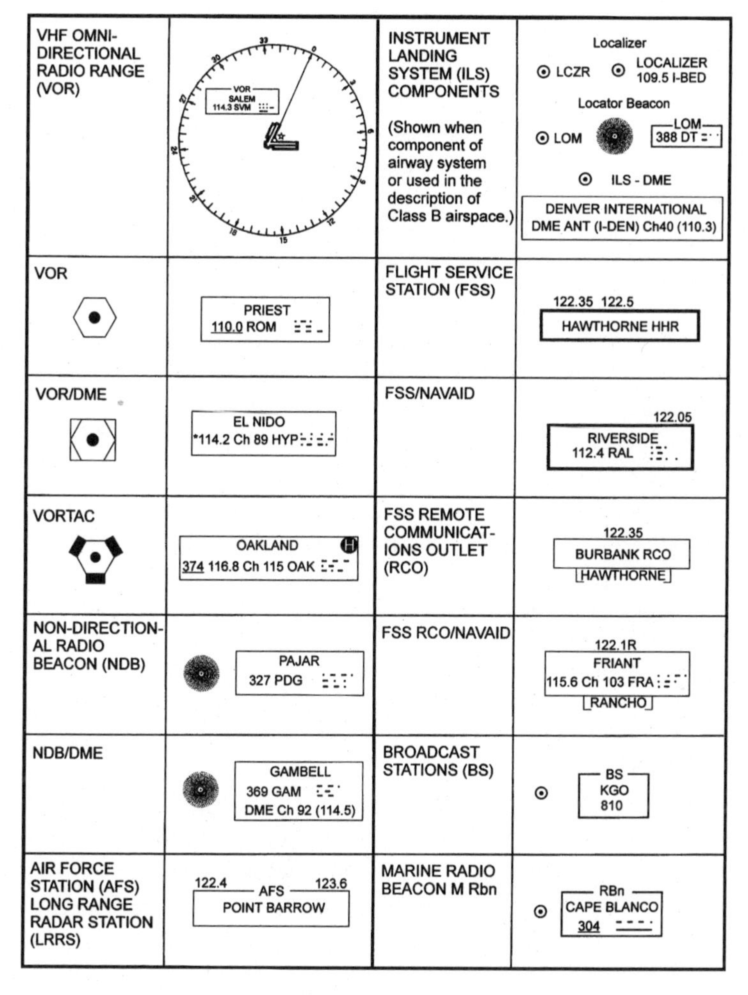

2-11 *NAVAID symbols provide type of facility and frequency.*

distance information from the DME. When the NAVAID is located on an airport, as shown in Fig. 2-11, the type of facility—in this case VOR only—appears above the NAVAID box. A circle and dot symbol indicate the location of the NAVAID on the airport. Otherwise, the appropriate symbol shown in Fig. 2-11 indicates the type of facility (VOR, VOR/DME, or collocated VOR and TACAN—VORTAC).

On charts, VHF and UHF NAVAIDs are shown in blue. Low- and medium-frequency beacons are shown in magenta, except that marine radio beacons are displayed in blue.

Normally a compass rose, oriented to magnetic north, accompanies a VHF NAVAID. An exception would be where closely located NAVAIDs would unduly clutter the chart. (A compass rosette is shown in areas void of VOR roses. Like the VOR rose, the compass rosette will be oriented to local magnetic variation.)

Location Identifiers

Locations were abbreviated with two letters in the early days of aviation, for instance NK was Newark. As the number of NAVAIDs and airports increased, three letter identifiers came into use. All VOR, VOR/DME, VORTAC, and many low-frequency radio beacons within the National Airspace System have three-letter identifiers. International identifiers consist of a four letter group. For example, KSFO (San Francisco, California, United States).

NAVAID identification boxes normally contain the name of the NAVAID, its frequency, three-letter identification, and the Morse code for that identification. Refer to the example of the Priest VOR in Fig. 2-11. Its frequency is 110.0 MHz. Note that the frequency is underlined. An underlined frequency indicates that no voice communications are available through the VOR. Priest's three-letter identifier is ROM. (Yes, the people responsible for assigning these identifiers do, on occasion, have a sense of humor.) Next is the Morse code for ROM: ·–· – – – – –.

Fact

With the exception of the United States Navy, NAVAIDs are named for an adjacent city or town. Their identities, to the extent possible, come from the city or town name. For example, Oakland-OAK. Originally, NAVAIDs not located on an airport used the identification of an adjacent airport. With the computerization of the air traffic system and aeronautical navigation, NAVAIDs with the same name but not physically located at the airport with the same IDENT had the potential for confusion. To avoid any misunderstanding for both pilots and controllers, these facility names have been changed over the last 20 years. For example, at one time the Long Beach, California, VOR, located 5 miles from the airport, name was changed to Seal Beach.

The next example in Fig. 2-11 is the El Nido VOR/DME. The NAVAID frequency is preceded with an asterisk (*). This means the NAVAID operates less than continuously or only on request. The TACAN channel follows the VHF VOR frequency. In the example Ch89, military aircraft would select DME channel 89 to receive distance information only on UHF. Civilian pilots would also receive distance information on DME channel 89. However, for simplification and to avoid confusion, civilian pilots select 114.2 on their DME receiver. This is accomplished through a system of paired frequencies. Every VOR on frequency 114.2 that has DME capability uses DME channel 89. The frequency pairing plan can found in the *Airport/Facility Directory*.

The next example in Fig. 2-11 is the Oakland VORTAC. Where a low-frequency (LF) beacon is colocated with a VHF NAVAID, the LF frequency may appear in the NAVAID box. The frequency of the LF facility is 374 kHz. Since this is an LF facility, it would be shown in magenta, rather than blue. Again, the underline indicates no voice communication over this frequency. Since this is a VORTAC, military pilots have access to both direction and distance information on TACAN channel 115. The letter H in the black circle means that Hazardous Inflight Weather Advisory Service (HIWAS) is available on the VOR frequency. (HIWAS is a continuous broadcast of hazardous weather conditions over selected NAVAIDs.) The letter T within the circle indicates a Transcribed Weather Broadcast (TWEB) is available over the VOR—there are only two such facilities left in the contiguous United States. The letter A means an automated weather observation is broadcast over the VOR frequency.

Nondirectional radio beacons (NDB) are shown in magenta and use the same frequency box information as VHF facilities. At some military locations DME is colocated with an NDB. In such cases the DME channel is provided in the frequency box. Additionally, the paired VHF frequency (114.5) is included. (Occasionally the military may use a UHF NDB or colocate a UHF NDB and DME. In these cases, the symbol is shown in blue, rather than magenta.

Communication Frequencies

Because of an ever-increasing need for more frequencies, the FAA has over the years reduced the spacing between communication channels. At present aviation radio frequencies are separated by 25 kHz, with the exception of the emergency frequency 121.5 MHz which still retains 100-kHz separation from adjacent channels. To accommodate 25-kHz spacing, VHF frequencies now go to three digits beyond the decimal point. For example, 133.275. To reduce charting space and communication time, by convention, the last digit of the frequency may be omitted and zeros are not shown or spoken. Frequency 133.275 may appear as 133.27 and spoken as one, three, three, point, two, seven; 122.000 appears as 122.0 and spoken as one, two, two, point, zero.

Heavy-line facility boxes indicate standard flight service station (FSS) communication frequencies (121.5 and 122.2 MHz, which are simplex—the pilot transmits and receives on the same frequency. Other FSS frequencies are printed above the box. For example, 123.6 for local airport advisory and FSS discrete frequencies. Routine communications should be accomplished on the station's discrete frequency. These frequencies are unique to individual facilities and locations. Their use will usually avoid frequency congestion with aircraft calling adjacent stations.

In Fig. 2-11 the Hawthorne (HHR) FSS has standard frequencies of 122.2 and 121.5 MHz. Additionally, at the same location 122.35 and 122.5 are available. The Riverside (RAL) FSS also has standard frequencies, as indicated by the heavy-line box. In addition to these frequencies, 122.05 is available at the same location.

FSS remote communications outlets (RCO) are illustrated in Fig. 2-11. The first example is the Burbank RCO. The associated FSS is Hawthorne, shown below the facility box. Above the facility box is the discrete frequency of 122.35. If a frequency is followed by the letter R (122.1R), the FSS has only receive capability on that frequency; therefore, the pilot transmits on 122.1 (or other displayed frequency). The pilot must tune another frequency, usually the associated VOR, to receive communications from the FSS. This duplex communication requires the pilot to ensure that the volume is turned up on the VOR receiver. For example, in Fig. 2-11 the Rancho FSS has a receiver located at the Friant VOR site on 122.1R, noted above the NAVAID box. A pilot wishing to communicate through the VOR would tune the transmitter to 122.1, and select Friant, 115.6, on the VOR receiver.

An FSS can transmit on many frequencies (VORs and remote outlets, for instance). With FSS consolidation, it is important for the pilot to advise the FSS which frequency is being monitored in the aircraft and the aircraft's general location. For example, "Reno radio, Cessna four three three four echo, listening one, two, two, point, six, Ely, over."

Note that only selected frequencies are depicted on aeronautical charts. Because enroute flight advisory service (flight watch) has a common frequency of 122.0, the frequency is not shown. Pilots calling flight watch should always include their approximate location on initial contact. Approach control and air route traffic control center frequencies are also omitted. Other frequencies are available on chart end panels or margins, or can be found in the *Airport/Facility Directory,* or obtained from an FSS.

Airspace

Controlled airspace serves three major purposes: establish weather minimums, impose minimum pilot qualifications and aircraft equipment, and warn of hazardous activities or protect national interests. Weather minimums allow enough ceiling, visibility, and cloud clearance for both VFR and IFR aircraft to "see and avoid." VFR weather minimums—and that's what they are, minimums—evolved in much the same way as controlled airspace. VFR weather minimums, especially below 10,000 ft, are much the same as they were in the beginning days of the Piper Cub and DC-3. With today's higher-performance airplanes minimum does not necessarily mean safe! In addition to weather minimums, Class D,

C, B, and A airspace impose one or all of the following requirements:

- Communications
- Aircraft equipment
- ATC clearance
- Minimum pilot qualifications

Special-use airspace (SUA) consists of airspace where activities must be confined because they pose a hazard to aircraft operations. These are designated as prohibited, restricted, warning, alert, military operations areas, and military training routes. Aircraft operations are prohibited within prohibited areas. These areas are established for security or other reasons associated with the national welfare. Aircraft operations are prohibited within Restricted Areas when the area is active. Restricted Areas are established for unusual, often invisible, hazardous activities, such as artillery firing, aerial gunnery practice, or guided missile firing. Warning areas are established for the same hazards as restricted areas, but over international waters. Alert areas inform nonparticipating pilots of areas that might contain a high volume of training or unusual activity. Pilots should exercise extra caution within these areas. Military operation areas alert pilots to military training activities. In addition to possible high concentration of aircraft, military pilots might conduct acrobatic flight and operate at speeds in excess of 250 knots below 10,000 ft. High-speed, low-level military operations are conducted along military training routes.

Ironically, controlled airspace affects the VFR pilot to a much greater degree than the IFR pilot. The reasons why airspace has become so complicated can be explained by its evolution. In the early days of aviation, all flying was visual; there was no such thing as controlled airspace or air traffic control. It wasn't until the mid-1930s that blind flying using instruments became practical. Then, after World War II, with the development of modern navigational aids and communications, instrument flying came into its own.

The purpose of controlled airspace is to provide a safe environment for instrument operations. Controlled airspace established weather minimums. As one might expect, controlled airspace originally developed around airports where air traffic was congested. The next logical extension included the then new electronic airway system. As radios become more common, certain airspace required the pilot to establish radio communications with the controlling authority. As jet travel increased, all airspace in the contiguous United States above 14,500 ft became controlled and flights above 24,000 ft required an IFR clearance. In the 1960s and 1970s, more and more airspace became controlled. The airspace where all aircraft were required to have a clearance was lowered to 18,000 ft, along with the airspace around major terminals. Specific communications and aircraft equipment requirements were established around smaller terminals.

Today, in addition to contending with various weather minimums in an alphabet soup of controlled airspace, the VFR pilot must establish communications in certain areas and ensure the aircraft has the required electronic equipment.

For our purposes, airspace on visual charts can be divided into two basic categories: controlled airspace and special-use airspace. Controlled airspace designated on visual charts is Class E, Class D, Class C, and Class B airspace. Any airspace not designated as one of the above is Class G—or uncontrolled—airspace. The primary purpose of controlled airspace is to protect IFR aircraft when weather conditions do not allow see-and-avoid separation. Class A airspace is not depicted on visual aeronautical charts.

By convention, when the ceiling and floor of controlled airspace are designated, the ceiling is in hundreds of feet MSL above a horizontal line, and the floor in hundreds of feet MSL below the horizontal line. This is illustrated in Fig. 2-12. In the example, the ceiling is 8000 ft MSL, with the floor 4000 ft MSL. In congested areas, different classes of airspace may overlap. In the example, the floor of Class C airspace is at the surface (SFC), and its ceiling is at the base of the overlying Class B airspace. [Class B airspace was originally designated as a terminal control area (TCA), thus the letter T indicating the base of the overlying airspace.]

Designated controlled airspace is considered to be active continuously, unless otherwise noted. In such cases pilots

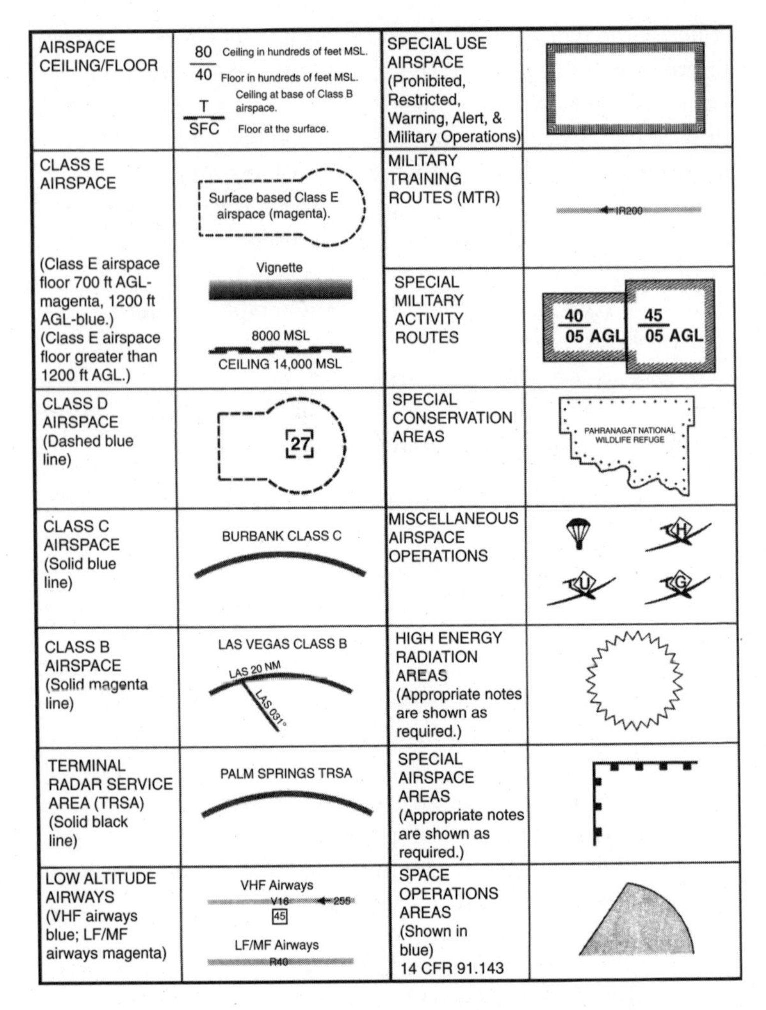

2-12 *Visual charts must provide enough airspace information to allow the VFR pilot to operate safely within an increasingly complex system.*

can expect to see a note: "See NOTAMs/Directory for Class D/E (sfc) eff hrs."

Class E airspace may begin at the surface, 700 ft AGL, 1200 ft AGL, or higher. Unless designated at a lower altitude, Class E airspace over the contiguous United States terminates at 14,500 ft MSL, except that area within 1500 ft of the surface—since this covers the whole country, it is not charted. (A sage pilot once observed that the first word in the regulations was except.) The ceiling of Class E airspace is at the floor of Class A airspace—18,000 ft. Refer to Fig. 2-12. Surfaced-based Class E airspace is depicted as a magenta dashed line. A magenta or blue vignette designates Class E airspace floors of 700 ft AGL and 1200 ft AGL respectively. The dark portion of the vignette shows the lateral extent of Class E airspace, with the lighter portion extending into the Class E area. A staggered thick blue line indicates a change in the floor of Class E airspace. With no altitude specified, Class E airspace begins at 1200 AGL. On one side of the staggered line the floor of Class E airspace will be specified in feet MSL—8000 MSL. When the ceiling is less than 18,000 ft MSL, the ceiling of Class E airspace is indicated—CEILING 14,000 MSL.

Class E airspace is probably the most confusing because its floor can vary from the surface to above 14,500 ft. To relate Class E airspace to the chart, refer back to Fig. 2-1. Let's begin in the lower left corner of Fig. 2-1 with the Taos airport. Surrounding the airport is a magenta vignette. The dark, sharp edge of the vignette is away from the airport and the lighter, feathered edge is oriented toward the airport. This means that the airspace surrounding the airport is Class E with a floor of 700 ft AGL. If we move north, toward the maximum elevation figure, we notice a blue vignette surrounding the airway that goes north from the Taos VOR. Since the lighter, feathered edge is oriented toward the airway, the floor of Class E airspace protecting the airway begins at 1200 ft AGL.

Now notice that just west of the maximum elevation figure is a staggered thick blue line—just south of San Antonio MTN—indicating a change in the floor of Class E airspace. As noted on the chart, the floor of Class E airspace begins at 12,000 ft MSL. There is no such symbology associated with the blue vignette on the east side of the airway. Therefore, the floor of Class E airspace begins at 1200 ft AGL. Recall the critical elevation example of 14,047 ft. Since Class E excludes that airspace within 1500 ft of the surface, over this mountain the floor of Class E airspace is at 15,547 ft.

All Class D airspace is surface based. Refer to Fig. 2-12. The lateral limits of Class D airspace are indicated by a blue dashed line. The upper limit is normally 2500 ft AGL. On charts the upper limit is designated by the blue number in hundreds of feet MSL in the dashed blue box. In the example the top of Class D airspace is 2700 ft MSL. A minus sign in front of the figure indicates Class D airspace extends from the surface to, but not including, the designated altitude. For example, −20 means from the surface to, but not including, 2000 ft MSL. Many Class D areas are effective part-time. A note on the chart will indicate effective hours and refer the pilot to the *Airport/Facility Directory* for the class of airspace (E or G) when Class D is not effective. (Many Class D areas have Class E surface extensions. When an extension is designed as Class E, is it depicted as a dashed magenta line.)

Class C airspace extends generally from the surface to 4000 ft AGL around airports with control towers and served by a radar approach control. The boundaries of Class C airspace are individually tailored on the basis of terrain and operational requirements. Class C airspace is charted using solid magenta lines. Various shelves exist beyond the surface-based airspace. Bases and tops are indicated in magenta using standard airspace ceiling/floor designations shown in Fig. 2-12. The appropriate approach control facility and frequencies are plotted with a magenta box.

Class B airspace surrounds the busiest airports. It generally consists of the airspace from the surface to 10,000 ft MSL with various shelves, sometimes referred to as an "upside-down wedding cake." Class B airspace is charted by using solid blue lines. Boundaries are defined by VOR radials, DME arcs, and prominent landmarks. As in Class C airspace, the bases and tops are charted. Bases and tops are indicated in blue with standard airspace ceiling/floor designations, shown in Fig. 2-12. The appropriate approach control facility and frequencies are plotted with a blue box.

Depicted in magenta and associated with Class B airspace is the 30-nm mode C arc. An altitude-encoding transponder is required within this airspace in accordance with 14 CFR 91.215. Appropriate notes as shown as required.

Terminal radar service areas (TRSA) designate airspace where traffic advisories, vectoring, sequencing, and separation of VFR aircraft are provided. TRSAs are designated Stage I, Stage II, and Stage III, which specify radar services that are available. The type of TRSA (Stage I, II, or III) can be found in the *Airport/Facility Directory*. TRSAs are shown by a solid black line.

Airways are shown in blue when established by a VHF NAVAID and magenta when established by an LF or MF facility. VHF airways are designated as V, or victor, airways below 18,000 ft. In Fig. 2-12 the example shows "victor sixteen" (V16). Appropriate radials are shown, as well as distances between NAVAIDs—in the example the distance between NAVAIDs is 45 nm. [On aeronautical charts all radials are magnetic and mileages are nautical (nm)]. Special use airspace is designated as prohibited, restricted, warning, alert, and military operations areas (MOA) and military training routes (MTR). Only the airspace effective below 18,000 ft MSL is shown. Space permitting, the type of area is spelled out, otherwise the type is abbreviated (P-56, R-6401, W-518). Details on the activity are provided in the margins of the chart. Figure 2-12 illustrates how these areas are designated on aeronautical charts. Areas are shown in blue, except MOAs, which are printed in magenta. MTRs are shown in black and labeled with the route number. Special military activity routes are shown in black with appropriate notes about the activity. MTRs where operations are conducted IFR are designated as IR, VFR operations VR. IR and VR routes operated at or below 1500-ft AGL will be identified by four-digit numbers (IR 1007, VR 1009). Operations conducted above 1500-ft AGL are identified by three-digit numbers (IR 205, VR 257). Route widths vary for each MTR and can extend several miles on either side of the charted centerline. Detailed route width information is available in the Flight Information Publication (FLIP) Ap/1B—a Department of Defense (DOD) publication. Special military activity routes alert pilots to areas where cruise missile tests are conducted.

Special conservation areas are designated with the symbology illustrated in Fig. 2-12. Landing, except in an emergency, is prohibited on lands or waters administered by the National Park Service, United States Fish and Wildlife Service, and United States Forest Service without authorization. All pilots are requested to maintain an altitude of at least 2000 ft above these areas.

Miscellaneous airspace operations consist of parachute jumps and glider, ultralight, and hang glider activity. These symbols are depicted in magenta. The specific type of activity is shown as follows: G—glider, U—ultralight, and H—hang glider. Where these symbols appear, pilots cannot expect to be alerted to the activity through NOTAMs. Details on these operations are contained in the *Airport/Facility Directory*.

National security areas are depicted as required in magenta. The reason for the area and affected altitudes are shown.

On sectional charts, the outline of terminal area charts (TACs) and chart insets are depicted as white lines.

Class A airspace consists of that area from 18,000 to 60,000 ft. What about the airspace above 60,000 feet? Well, it's Class E airspace. To help visualize the vertical extent of airspace, refer to Fig. 2-13. Basic VFR weather minimums are also shown.

What happened to Class F airspace? Class F is an International Civil Aviation Organization (ICAO) airspace classifica-

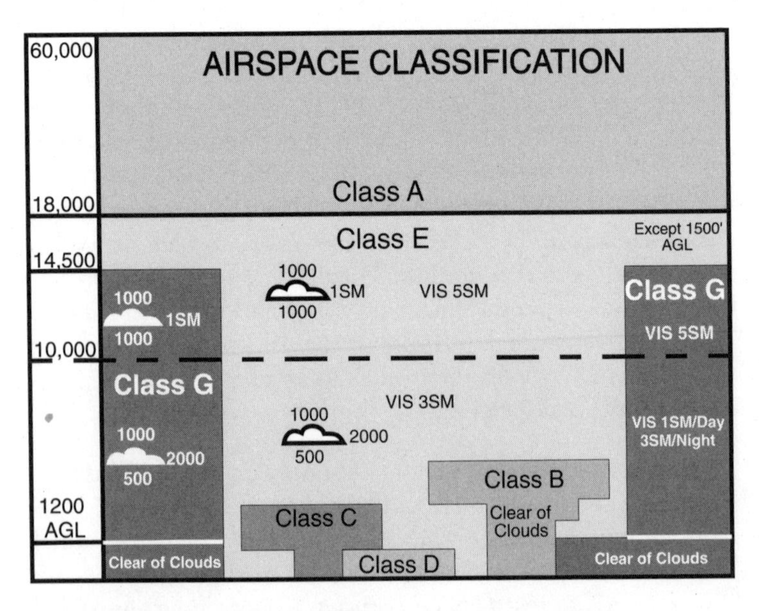

2-13 *Airspace imposes weather minimums and may establish minimum aircraft equipment and pilot qualifications.*

tion. IFR and VFR flights are permitted. Air traffic advisory service and flight information service is provided on request. An ATC clearance is not required. This category is not used in the United States.

Fixed-wing special VFR is normally available in surface-based controlled airspace. In certain high-density surface-based airspace, special VFR is prohibited. This is indicated in the airport data block by NO SVFR.

Navigational information

Navigational information consists of isogonic lines and values, airway intersection depictions, aeronautical and marine lights, visual ground signs, and VFR checkpoints.

Isogonic lines (lines of equal magnetic declination for a given time) provide the pilot with the difference between true north and magnetic north. These lines, printed in magenta, show values east (E) and west (W) of the igonic line—the line of zero declination. In Fig. 2-14, the example declination is 42°E. (Recall that when converting true values of course or heading to magnetic subtract easterly variation and add westerly variation. "East is least and west is best.") These lines and values are updated every 5 years. Local magnetic notes alert pilots to areas of magnetic disturbance where the magnetic compass might be unreliable, often because of large deposits of iron ore.

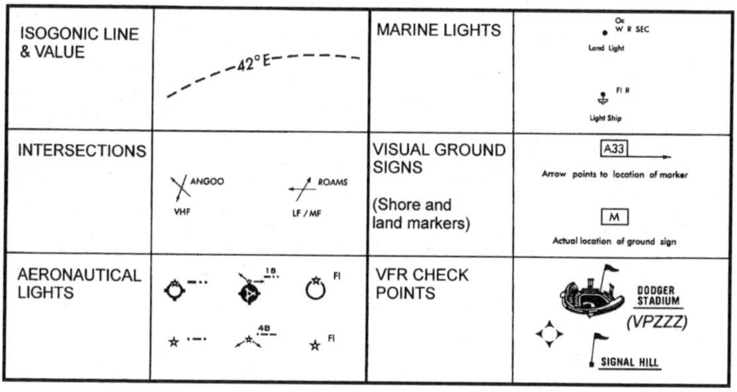

ISOGONIC LINE & VALUE	42°E	MARINE LIGHTS	Oc W R SEC Land Light Fl R Light Ship
INTERSECTIONS	ANGOO VHF ROAMS LF / MF	VISUAL GROUND SIGNS (Shore and land markers)	A33 Arrow points to location of marker M Actual location of ground sign
AERONAUTICAL LIGHTS	18 Fl 48 Fl	VFR CHECK POINTS	DODGER STADIUM (VPZZZ) SIGNAL HILL

2-14 *Navigational and procedural information includes lines of magnetic variation and aeronautical lights.*

Magnetic Jerk

The inner core of the earth is solid and surrounded by a molten outer core. The inner, solid core actually rotates at a slightly faster rate than the rest of the planet. Over the period of a year, the inner core rotates about 1 degree more than the earth's surface. The earth's magnetic field is always slowly changing. That's why declination is updated every 5 years. In 1912 and again in 1969 the earth's magnetic field appears to have gone through a minijerk to a new alignment.

It's been known for quite some time that the earth's magnetic field has changed polarity from time to time. In other words, your aircraft's compass, which now points to magnetic north, would point to magnetic south. The big switch, however, usually takes about 4500 years. Hopefully, we won't be faced with a "big jerk" in the near future.

Named intersections that can be used as reporting points are depicted on some visual charts. Their depiction is illustrated in Fig. 2-14. Intersections depicted in blue are established with VHF NANAIDS and those using LF/MF NAVAIDS are shown in magenta. The intersections consist of a five letter name that might be difficult to pronounce. Airway radials are depicted; however, unfortunately, the cross fix radial may not be shown. Therefore, the use of airway radials is limited.

Aeronautical lights at one time were a primary means of navigation. For example, in the 1930s lighted airways were a primary means of navigation. Light symbols are depicted on visual charts for their navigational value; however, because electronic aids have taken over the majority of navigational needs, many of the old large airport beacons have been replaced with smaller units. This can lead to confusion.

Case Study

The Bakersfield, California, Meadows Airport has one of the small, less-intense beacons, while the nearby Shafter Airport has the older, large unit. Pilots, including Army helicopter pilots, regularly key on the Shafter beacon, and even land at the wrong airport. The larger lights can often be seen twice as far as the smaller units.

Recall that airport rotating or oscillating beacons are indicated by a star adjacent to the airport symbol and normally operate sunset to sunrise. Rotating lights with flashing code identification and course lights are relics from the lighted airway days, and almost all have been decommissioned. A few still remain and their symbols are depicted in Fig. 2-14.

The code beacon, which can be seen from all directions, is used to identify the stations. The beacon flashes the three or four character identifier. The two symbols on the right in Fig. 2-14, aeronautical lights, illustrate the depiction of a rotating light with flashing code identification. Course lights, which can be seen clearly from only one direction, are aligned with the lighted airway. These are represented by the center two symbols in Fig. 2-14, rotating light with course lights and site number. The right two symbols in Fig. 2-14 are flashing lights only.

For their navigation value, marine lights, with the characteristics of the light, are shown on visual charts, as depicted in Fig. 2-14. Marine lights are white, and alternating lights are red and white, unless otherwise noted. Table 2-1 contains marine light characteristics.

Table 2-1. Marine light characteristics

R	Red	W	White
G	Green	B	Blue
SEC	Sector	F	Fixed
Oc	Single occulting	Oc (2)	Group occulting
Oc (2+1)	Composite group occulting	Iso	Isophase
Fl	Flashing	Fl (2)	Group flashing
Fl (2+1)	Composite group flashing	Q	Quick
IQ	Interrupted quick	Mo (A)	Morse code
FFl	Fixed and flashing	Al	Alternating
Gp	Group		

Below is an explanation of the basic light signatures that are not apparent.

- *Occulting.* An intermittent light that is on longer than off.
- *Isophase.* A light that flashes at equal intervals.
- *Composite group.* A flashing light whose flashes are combined in alternating groups of different numbers.
- *Quick.* A light that flashes 60 times or more a minute.
- *Interrupted quick.* A light in which 5 seconds of quick flashes is followed by 5 seconds of darkness.

Visual ground signs that are easily recognizable from the air are depicted. These are often in the form of a letter obtained from the first letter of a nearby town. Most visual signs are left over from the early days of aviation, have faded, and are of little value; others are large and prominent.

Case Study

A chart that I used in England contained a large picture of a horse. Overflying the area, I was a little surprised to see on the ground an extremely large, perfectly proportioned steed, just as depicted on the chart.

Certain cultural features that can be clearly seen from the air are used as VFR checkpoints. These checkpoints are designated with a magenta pennant, as shown in Fig. 2-14. Prominent VFR checkpoints are depicted with pictorial symbols (black) with elevation data (where available) in blue, such as the prominent Golden Gate Bridge or Dodger Stadium, as illustrated in Fig. 2-14. In Fig. 2-14 the four-pointed symbol indicates a GPS waypoint. When a VFR checkpoint is stored within the GPS or area navigation data base its identification is shown in magenta (*VPZZZ*).

3

Standard visual charts

World War I stimulated interest in aeronautical charts in the United States. Lawrence Sperry, of Sperry Gyroscope Company, had prepared an aeronautic map of Long Island in 1917. The map was presented to Rear Admiral Robert E. Peary, chairman of the Committee on Aeronautic Maps and Landing Places of the Aero Club of America.

Commercial aviation in the United States was launched on August 12, 1918, with the initiation of regular airmail service between Washington and New York. Pilots were forced to use railroad and road maps, or pages from atlases. As late as 1921, with transcontinental airmail operations day and night, no aeronautical charts existed. Pilots noted times and course between prominent landmarks. If they were lucky they flew two trips behind veteran pilots; if not, just one trip. Notes from various pilots were assembled and published by the Post Office Department. These *Pilot's Directions* contained distances, landmarks, compass courses, and emergency landing fields, with services and communications facilities at principal points along the route.

With the passage of the Air Commerce Act in 1926, the Department of Commerce became responsible for the production of aeronautical charts for the nation's airways. The first chart published in 1927 was a strip map that covered the air route from Kansas City to Moline, Ill. These early charts depicted prominent topographical features for visual flying and contained the locations of the newly installed airway lighted beacon system for night operations. The strip map concept was extended to other lighted airways between major airports throughout the late '20s. Figure 3-1 illustrates a strip map for the route Milford to Salt Lake City, Utah, circa 1929.

3-1
These early charts depicted prominent topographical features for visual flying and contained the locations of the newly installed airway lighted-beacon system for night operations.

With more flying conducted away from established airways, it became apparent that a system of charts was needed to provide complete coverage. Recommendations of the Committee on Aerial Navigation Maps in 1929 prompted the Coast and Geodetic Survey to develop a series of 92 sectional aeronautical charts for the United States.

World War II compressed a quarter-century of peacetime aeronautical chart development into a few years. The need for all types of charts was urgent and insatiable. The term *aeronautical chart* became firmly established during this time. Previously, charts were referred to as air maps, aeronautical maps, flight maps, or aeronautical charts. Most charts were variations of those in existence prior to 1939, which was a time saver. Charts rolled off the presses by the millions.

Sectional and world aeronautical charts evolved through several stages in the 1960s and 1970s. Mostly because of economic considerations, charts were printed on both sides, reducing the total number of charts but unfortunately eliminating what many pilots considered useful information printed on the reverse side. Much of this information was transferred to various sections of the then *Airman's Information Manual* (AIM).

In 1992 the National Oceanic and Atmospheric Administration proposed the elimination of world aeronautical charts (WACs). The government wanted pilots to switch to Defense Mapping Agency Operational Navigation Charts (ONCs). Various aviation organizations, citing safety and pilot demand, lobbied for the continuation of the WAC series. Fortunately, the government listened and this series is still available.

Implementation of complex airspace configurations around major airports, Class B airspace, fostered development of the terminal area chart (TAC) for improved presentation of Class B dimensions and better resolution of ground references.

Case Study

Prior to implementation of Class B airspace around the Los Angeles International airport, I was on several occasions almost nailed by air carrier jets over Santa

Monica. Class B airspace has allowed a much safer flight for general aviation aircraft by mandating that the jets maintain a higher altitude in this area and remain within Class B airspace.

On October 1, 2000, responsibility for aeronautical charts was moved from the Department of Commerce (DOC), National Ocean Service (NOS), to the FAA's National Aeronautical Charting Office (NACO). NACO publishes a number of aeronautical charts specifically designed to assist the pilot with visual or VFR navigation: planning charts, sectional charts, terminal area charts (TACs), and world aeronautical charts (WACs). Interestingly, these charts are often more complex than those used for instrument or IFR navigation. Chapter 2 discussed the dozen or so types of airspace plus a description of terminology and symbols common to visual charts; therefore, in this chapter only those terms and symbols unique to individual charts are presented.

Case Study

Pilots had previously paid a nominal fee for charts; however, the government in the face of increasing deficits decided pilots should pick up more of the tab, and prices for charts and publications skyrocketed. At one point it was proposed that users pay the development as well as printing costs, which would have put the price of charts almost out of sight for many pilots. Fortunately, aviation organizations, notably the Aircraft Owners and Pilots Association (AOPA), pressured the government into a compromise. Many still think prices are outrageous, but considering the information available, and the cost of charts in other countries, we're still getting a bargain.

Planning charts

Planning charts, as the name implies, are designed for the initial portion of flight preparation, as opposed to operational charts—sectionals, TACs, and WACs used for flight planning and navigation. These charts are most useful for planning long trips, usually with several or more legs. For example, I have planned and executed two VFR crossings of the United States, one from California to New York and the other from California to Florida in a Cessna 150. Each trip was nothing more than a series of short cross-country flights. Here is where planning charts are of most value. Planning chart coverage is illustrated in Fig. 3-2.

IFR/VFR low-altitude planning chart

The United States IFR/VFR low-altitude planning chart combines many of the features previously found on the discontinued IFR/VFR wall planning chart and the flight case planning chart. The 1:3,400,000 (1 in equals 47 nm) scale chart is designed for preflight planning below 18,000 ft.

3-2 *Planning charts are designed for the initial portion of flight preparation.*

Revised annually, this five-color, two-sided chart may be obtained either flat (35 in × 36 in) or folded (5 in × 9 in). It has the eastern half of the United States on one side and the western half on the other side. For wall display, the purchase of two flat charts is required to show the conterminous United States. This chart is often found in planning rooms of flight schools and fixed-based operators (FBOs). Features include:

- Low altitude airways
- Navigational aids
- Airports—3000 ft paved or with an instrument approach
- Airspace classification
- Maximum elevation figures
- Time zones
- ARTCC boundaries
- Sectional chart outlines
- IFR enroute chart outlines
- Special-use airspace
- International and state boundaries
- Large bodies of water
- Selected cities and towns

Gulf of Mexico and Caribbean planning chart

The Gulf of Mexico and Caribbean planning chart has a still smaller scale than the IFR/VFR low-altitude planning chart. This chart is designed for preflight planning of flights through and around the Gulf of Mexico and Caribbean. Intended to be used in conjunction with world aeronautical charts, it is printed on the back of the Puerto Rico–Virgin Islands VFR terminal area chart. The area of coverage is shown in Fig. 3-2. The chart is 60 × 20 in, which can be folded to the standard 5 × 10, and has a scale of 1:6,192,178 (1 in equals 85 nm). Features include:

- Airports of entry
- Special use airspace below 18,000 ft
- Significant bodies of water
- International boundaries
- Large islands and island groups

- Capital cities and cities where an airport is located
- Selected other major cities
- Air mileage between airports of entry
- Index of world aeronautical charts
- Directory of airports, including facilities, servicing, and fuel
- Department of Defense requirements for civilian use of military airports
- Checklist for ditching
- Runway visual range (RVR) conversion table from feet to meters
- Emergency procedures

Charted VFR flyway planning charts

Charted VFR flyway planning charts are designed to assist pilots planning flights through or around high-density areas such as Class B and Class C airspace. These two color charts are printed on the back of selected terminal area charts (TACs) with coverage corresponding to the TACs. The following TACs contain VFR flyway planning charts:

- Atlanta
- Baltimore–Washington
- Charlotte
- Chicago
- Cincinnati
- Dallas–Fort Worth
- Denver
- Detroit
- Houston
- Las Vegas
- Los Angeles
- Miami
- Orlando
- Phoenix
- St. Louis
- Salt Lake City

- San Diego
- San Francisco
- Seattle

Charted VFR flyway planning charts, as the name implies, are not to be used for navigation, or a substitute for the TAC or sectional charts. Features include:

- Airports
- NAVAIDs
- Special-use airspace
- Class B airspace
- Class C airspace
- Class D airspace
- VFR flyways (suggested headings and altitudes)
- Procedural notes
- Military training routes
- Selected obstacles
- VFR checkpoints
- Hydrographic features
- Cultural features
- Terrain relief designated as VFR checkpoints
- Critical spot elevations

Symbols used on VFR Flyway Planning Charts are standard. In addition, as the chart name implies, recommended VFR flyways are depicted. These consist of heavy blue lines with directions, mileages, and altitudes. They may include latitude/longitude coordinates for navigational reference points—for use with area navigation equipment. More and more Global Positioning System (GPS) waypoints are being added to these charts to assist pilots with precise navigation. Standard IFR routes are also depicted. These are portrayed as dashed blue lines that include standard arrival and departure routes, altitudes, and appropriate notes as required.

VFR corridors have been established at a number of Class B airspace locations. These corridors allow VFR transition through Class B airspace. Some require an ATC clearance, while others have special communications frequencies and transponder codes. Before attempting to negotiate any of these corridors, pilots must be completely familiar with their geographical location, navigational procedures, and any special communications or clearance requirements.

Using planning charts

Recall that a long cross-country flight is nothing more than a series of smaller, individual legs. Take for example a flight from Van Nuys, California, to Jamestown, New York, and return. The first consideration is the aircraft, the second consideration is the pilot and passengers. It doesn't do much good to plan 4-hour legs with 2-hour bladders. In a Cessna 150, 250-mile legs are comfortable; in a Bonanza, depending upon fuel load, 350- to 400-mile legs would be comfortable; in a Hughes 269 helicopter, with its speed and limited fuel, only 100-mile legs are practical. It all depends on the aircraft, fuel, and pilot/passenger comfort.

With average legs in mind, we proceed to the planning chart, realize aircraft performance, and consider terrain, airports and services, and controlled and special-use airspace. I don't like flying over high, rough terrain, or unpopulated areas of deserts or swamps, if at all possible. Enroute destinations are selected for the services available. We can plan the flight to major airports, if we have the proper electronic equipment, or to uncontrolled fields, should we wish. We have to avoid prohibited and restricted areas and we might wish to avoid Military Operation Areas (MOAs).

The flight from California to New York was preplanned with routes, altitudes, alternates, and services considered. For example, Class B and C airspace, TRSA, control towers, and FSS frequencies were logged. Of 24 individual legs, only one, from Huntington, West Virginia, to Cincinnati, Ohio, was not on the original itinerary. The unscheduled stop was required to repair the radio.

From the planning chart we determine which sectional charts are needed. From these charts we can either plan to avoid Class B airspace or obtain required terminal area charts to negotiate or avoid their heavy concentrations of traffic.

Sectional charts

Sectional charts are designed for visual navigation of slow to medium-speed aircraft. These multicolored charts provide the most accurate means of pilotage—navigating the aircraft by means of ground reference—because of their scale. The chart is 20 × 60 in, which can be folded to the standard 5 × 10 in and has a scale of 1:500,000 (1 in equals 7 nm). They are revised semiannually, except in Alaska. Alaskan charts are revised annually, except for Point Barrow, Fairbanks, Anchorage, and Seward. Sectional charts are named for a major city within the area of coverage. Figure 3-3 contains sectional chart coverage for the contiguous United States, and Fig. 3-4 shows coverage for Alaska. In addition to those charts shown in Fig. 3-3 and Fig. 3-4, there are the Hawaiian Islands and Puerto Rico/Virgin Islands sectionals. Features include:

- Visual aids to navigation
- Radio aids to navigation
- Airports
- Controlled airspace
- Restricted areas
- Obstructions
- Topography
- Shaded relief
- Latitude and longitude lines
- Airways and fixes
- Other low-level-related data

The Hawaiian sectional is the only chart of this series that is not oriented to true north. This is necessary to portray all the islands of the group on a standard size sheet, which also contains the Mariana and Samoa Island groups.

Recall from Chap. 1 that before using any chart, or aeronautical publication, a pilot's first task is to determine currency. That means reviewing the cover page for effective and obsolescent dates. Figure 3-5 is the cover page and data panel from the San Francisco Sectional Aeronautical Chart. The chart is a Lambert conformal conic projection with standard parallels 33°20′ and 38°40′, based on the North American

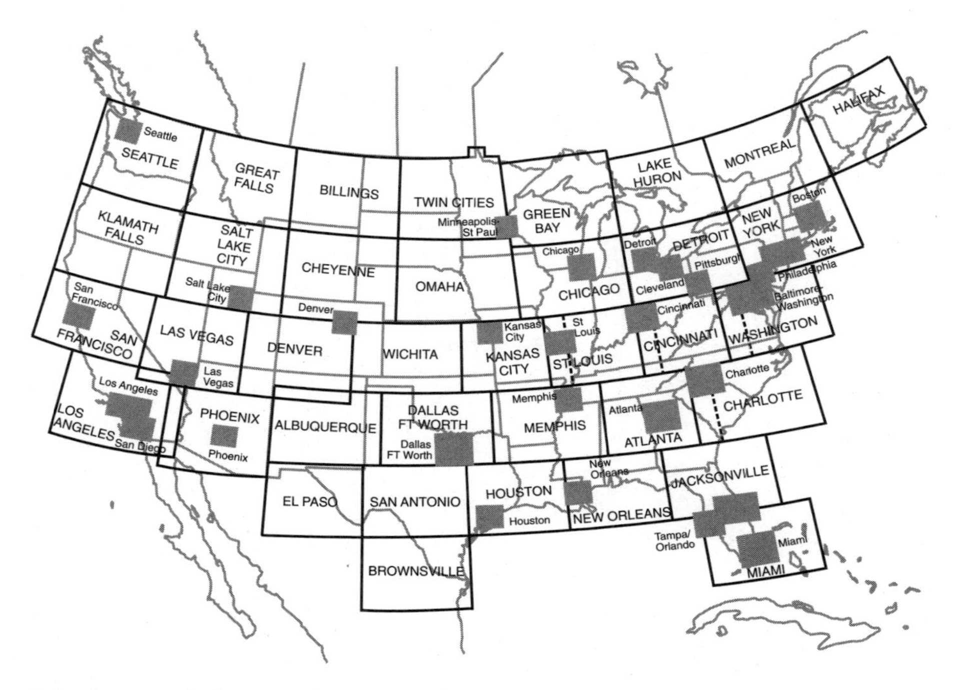

3-3 *Sectional charts are named for a major city within their area of coverage.*

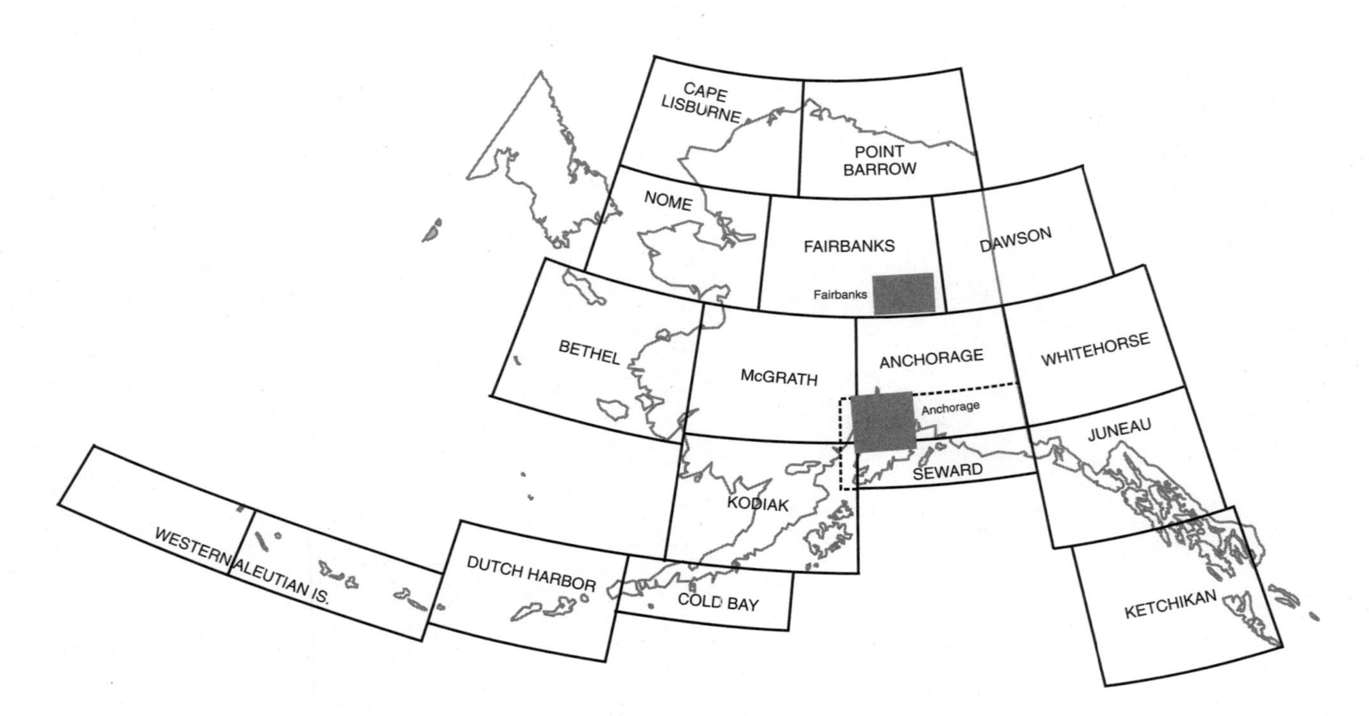

3-4 *Sectional charts are revised semiannually, except for most Alaskan charts, which are revised annually.*

SAN FRANCISCO
SECTIONAL AERONAUTICAL CHART
SCALE 1:500,000

Lambert Conformal Conic Projection Standard Parallels 33°20′ and 38°40′
Horizontal Datum: North American Datum of 1983 (World Geodetic System 1984)
67 TH EDITION September 6, 2001
Includes airspace amendments effective September 6, 2001
and all other aeronautical data received by July 12, 2001
Information on this chart will change; consolidated updates of chart changes are available every 56 days in the AIRPORT / FACILITY DIRECTORY (A/FD). Also consult appropriate NOTICES TO AIRMEN (NOTAMs) and other FLIGHT INFORMATION PUBLICATIONS (FLIPs) for the latest changes.
This chart will become *OBSOLETE FOR USE IN NAVIGATION* upon publication of the next edition scheduled for *MARCH 21, 2002*

PUBLISHED IN ACCORDANCE WITH INTERAGENCY AIR CARTOGRAPHIC COMMITTEE SPECIFICATIONS AND AGREEMENTS, APPROVED BY:
DEPARTMENT OF DEFENSE • FEDERAL AVIATION ADMINISTRATION

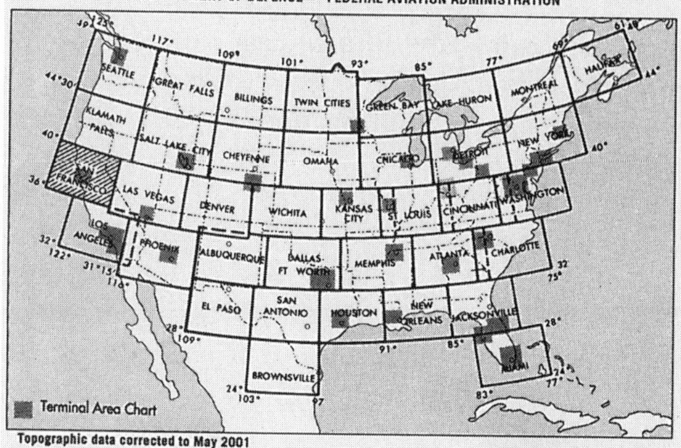

Topographic data corrected to May 2001

CONTOUR INTERVAL 500 feet
Intermediate contour 250 feet
Auxiliary contours 100 foot intervals
500 250 100

HIGHEST TERRAIN elevation is 14491 feet
located at 36°35′N – 118°18′W
Spot elevation 4254
Approximate elevation x3200
Doubtful locations are indicated by omission of the point locator (dot or "x")

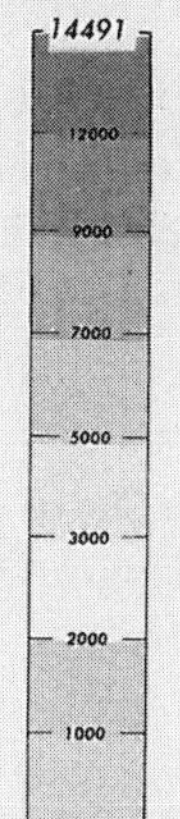

MILITARY TRAINING ROUTES (MTRs)

All IR and VR MTRs are shown, and may extend from the surface upwards. Only the route centerline, direction of flight along the route and the route designator are depicted – route widths and altitudes are not shown.

Since these routes are subject to change every 56 days, and the charts are reissued every 6 months, you are cautioned and advised to contact the nearest FSS for route dimensions and current status for those routes affecting your flight.

Routes with a change in the alignment of the charted route centerline will be indicated in the Aeronautical Chart Bulletin of the Airport/Facility Directory.

Military Pilots refer to Area Planning AP/1B Military Training Route North and South America for current routes.

CONVERSION OF ELEVATIONS

FEET (Thousands) 0 2 4 6 8 10 12 14 16 18 20 22 24 26 28 30
METERS (Thousands) 0 1 2 3 4 5 6 7 8 9

Published by the
U.S. Department of Transportation
Federal Aviation Administration
National Aeronautical Charting Office

3-5 *Chart cover pages have effective and obsolescence dates; margins contain data panels with supplemental information.*

Datum of 1983 (World Geodetic System 1984). This is the sixty-seventh edition of this chart, effective September 6, 2001. It includes airspace amendments effective on September 6, 2001 and all other aeronautical data received by July 12, 2001. It refers the user to the *Airport/Facility Directory* and NOTAMs for updates. Finally, it advises the user that the chart will be "...obsolete for use in navigation..." with the next edition scheduled for March 21, 2002.

The margins of sectional charts contain data panels with supplemental control tower information, selected radar approach control frequencies, and SUA information. For example, one entry advises pilots that the China Lake NWC (Naval Weapons Center) control tower operates between 6:30 A.M. and 10:30 P.M. Monday through Friday on frequencies 120.15 MHz for civilian aircraft and 340.2 MHz for military aircraft. Ground control (GND CON), automatic terminal information service (ATIS), and the availability of radar approaches are also noted. This information is followed by selected Class B, Class C, TRSA, and radar approach control facilities.

Data regarding special use airspace includes altitudes, effective times, and controlling agency. The control of some restricted areas are routinely released to VFR and IFR operations when not in use. Pilots can obtain this information from the local flight service station or controlling agency, usually an approach control or the respective air route traffic control center (ARTCC). The status of some restricted areas is routinely reported through NOTAM (D)s and will be part of a standard FSS or DUAT briefing.

Case Study

I queried a military airport tower about the status of a restricted area along my route. The controller advised that it was "cold." Later, in the middle of the restricted area, I contacted an approach controller. This controller advised me to remain clear of all restricted areas. It turned out the area was not in use, but would be in a few hours.

Military operation areas (MOAs) can be confusing. They are not restricted areas; their purpose is to alert pilots to the existence of potentially hazardous military operations. Pilots should exercise additional vigilance while transiting active MOAs. Flight service station controllers often receive requests for published MOA activity. Pilots should first refer to the chart for this information. If the MOA is activated by NOTAM, an automated FSS will be able to provide its status. In Alaska, nonautomated FSSs will be able to provide MOA status only within approximately 100 miles of the FSS; therefore, a pilot will have to check with FSSs enroute for MOA activity beyond this distance. As with restricted areas, MOA status can also be obtained from the respective controlling agency.

With preliminary planning completed, it's time to open the sectionals or WACs and prepare a navigation log with routes, true courses, distances, and magnetic variation, along with communication and navigation frequencies. This can be done on any one of many commercially available forms.

Planning a long trip several days, even several weeks, in advance is not a difficult chore. What about the weather? The general weather patterns of the United States are well documented. For example, we know about the winter storms of the Midwest and East, the convective weather of the Midwest in the spring and summer, the heat of the southwestern deserts in summer, and coastal low clouds of the Pacific states in late spring, summer, and fall. If needed, a call to the area's FSS will often provide the general weather conditions for a certain area and time of year. But, please, don't expect specifics. How can we use this general weather for flight planning?

Case Study

With a VFR-only Cessna 150, flying out of the Los Angeles Basin, I always plan to depart on the first leg in the afternoon, after the fog clears. This also permits flight over the desert in the late afternoon or early evening when the turbulence has diminished. Departures are then planned early the next morning to avoid desert convective activity and turbulence. Another solution is to move the airplane inland out of the affected coastal areas. Time of year is also important. If you plan winter operations, VFR only, be prepared for delays. May, June, September, and October seem to have the best flying weather.

What about winds aloft? If we don't try to stretch our trip legs, 10 to 20 knots of wind either way shouldn't present a problem.

Case Studies

We planned a leg from Phoenix to Albuquerque; however, because of the distance we could not tolerate any headwind component; therefore, we planned an alternate, using Gallup. If we were not on time at a specific point, about halfway, we would divert to our planned alternate. Fortunately, the winds were with us and we proceeded to our planned destination.

On another flight from Prescott to Albuquerque things just were not meant to be. Crossing Winslow, Arizona, the Cessna's ground speed never reached three digits. I changed the flight plan and preceded to Gallup. Hoping that a stronger than forecast head-wind will abate is folly.

Despite the best plans of mice and men, things go wrong. On a leg from Kalamazoo to Detroit, flight service advised of a thunderstorm over Jackson, Michigan. Further checking indicated that to the south, toward Toledo, was clear. A slight diversion and pilotage navigation took us safely to the new destination.

Study the charts in advance to determine best routes, comfortable legs, adequate services at destinations, and possible alternates. With everything planned, if a problem occurs, a pilot is in a much better position to evaluate the situation and develop a sound alternative.

Terminal area charts

Terminal area charts (TACs) replaced the local chart series beginning in the early 1970s. These multicolored charts depict Class B and Class C airspace, and TRSAs with much more detail than is available with sectional charts because of their larger scale. They are designed for pilots operating from airports within or near congested airspace or transiting the vicinity. Charts are 20 × 25 in, which can be folded to the standard 5 × 10 in, and have a scale of 1:250,000 (1 in equals 3 nm).

TAC charts are revised semiannually. Charts are named for the congested airspace they depict, and locations are shown in gray in Fig. 3-3 and Fig. 3-4. The Anchorage/Fairbanks TAC is based on the Anchorage Class C airspace and the Fairbanks TRSA. The Honolulu TAC is on the Hawaiian Sectional Chart. Features include:

- Visual aids to navigation
- Radio aids to navigation
- Airports
- Controlled airspace
- Restricted areas
- Obstructions
- Topography
- Shaded relief
- Latitude and longitude lines
- Airways and fixes
- Other low-level-related data

Improved scale allows for a great deal of topographical detail. Along with depicted NAVAIDs, TACs allow a pilot to safely navigate in the vicinity of, and remain clear of, congested airspace. Even with this detail, in marginal weather conditions new pilots might not have the experience to navigate in these areas. Remembering that minimums specified in the regulations are just that, minimums, not necessarily safe. Each pilot must set standards, based on training and experience. This might mean avoiding terminal airspace altogether, only flying in clear weather, or obtaining additional training from a qualified instructor. Every new pilot planning to fly into Class B airspace, or other congested airspace, should make at least one trip with an instructor or an experienced pilot.

Pilots should not necessarily avoid congested airspace because of its complexity. Properly trained pilots can, and do, safely negotiate Class B airspace. This typically requires good communications skills and attention to ATC instructions. General aviation pilots might want to adopt the airline policy of a "sterile cockpit"—that is, no unnecessary conversation or distractions when the aircraft is operating in these areas.

Case Studies

I had a primary student who wanted the experience of transiting the San Francisco Class B airspace on a Sunday afternoon. I had explained the limitation of this procedure. We contacted Bay Approach east of Hayward and requested direct Half Moon Bay at 4500 ft. I was astonished by their reply: "Cleared as requested." Well, we did encounter some route and altitude changes; but, we essentially did proceed as we had initially requested.

A few months later I made a similar request to Bay Approach. This time the controller responded: "Well, that isn't going to happen!" We were cleared through Class B airspace, ironically via almost the same route as our previous flight.

On a flight in Texas our direct route took us through the center of the Houston Class B airspace. I called approach over a VOR just prior to their Class B airspace, made my request, and with a few diversions for traffic, ATC was able to accommodate my request.

The previous examples are not to say that we will always get what we want. But, the point is that ATC is supposed to accommodate our requests if traffic allows. And, to the extent possible, most of the time they do.

Helicopter route charts

Helicopter route charts are designed primarily to depict helicopter routes in and around major metropolitan areas. Since the helicopter charts have a longer life span than other NACO products, all new editions are printed on a synthetic paper. The use of this paper will result in an increased cost, of course. Helicopter route charts are available for the following locations:

- Baltimore-Washington (District of Columbia)
- Boston

- Chicago
- Houston (includes Galveston Bay)
- Los Angeles
- New York (downtown Manhattan, Statue of Liberty/Ellis Island, Weschester)

Scale is the same as a TAC's and dimensions are similar, except for the New York chart, which includes a larger-scale inset of Lower Manhattan, and the Hudson and East Rivers, the Boston chart with its downtown Boston inset, and Chicago with its O'Hare and vicinity inset. Charts contain specific route descriptions as illustrated in Fig. 3-6. The inset in Fig. 3-6 contains symbols unique to this chart. In addition to specific helicopter routes, features include:

- Pictorial symbols of prominent landmarks
- Public, private, and hospital heliports
- NAVAID and communications frequencies
- Selected obstructions
- Roads
- Spot elevations
- Commercial broadcast stations
- Class B, C, and D airspace boundaries

Large metropolitan areas without published helicopter route charts often have local procedures that accomplish the same purpose. Letters of agreement between air traffic control facilities in these areas designate helicopter checkpoints, routes, and route names. Pilots planning operations in these areas should contact local pilots or ATC facilities for details on these procedures. This might require the pilot or operator to become a signatory to the letter of agreement, stating that he or she understands and will comply with its provisions.

World aeronautical charts

World aeronautical charts (WACs) are designed for visual navigation by moderate-speed aircraft and aircraft operating at higher altitudes, up to 17,500 ft MSL. Because of their smaller scale, these charts cannot show the detail of sectionals and

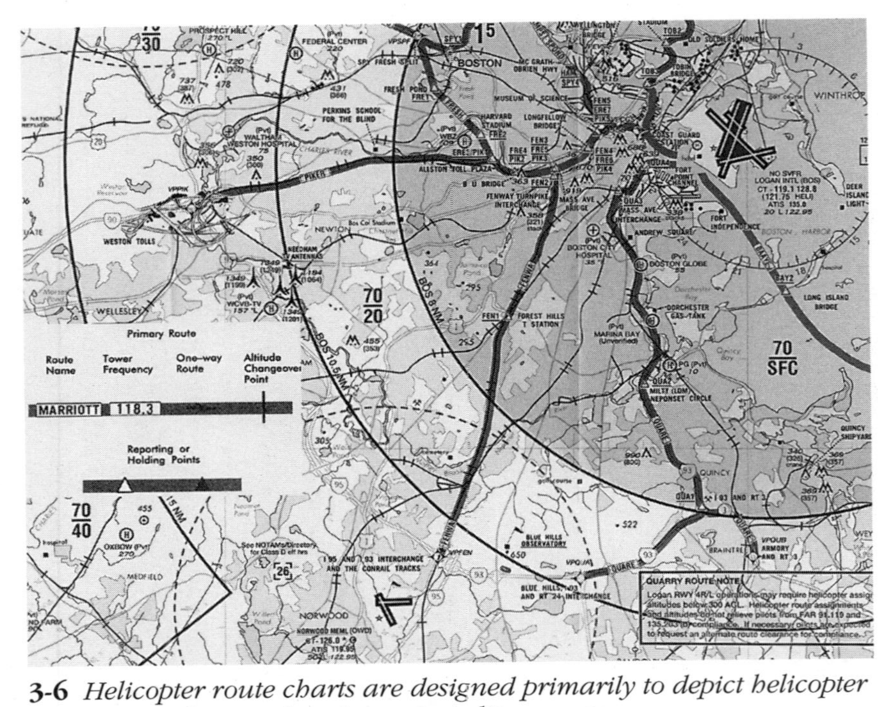

3-6 *Helicopter route charts are designed primarily to depict helicopter routes in and around major metropolitan areas.*

TACs, for example, the limits of Class D and Class E airspace. WACs are normally not recommended for student or new pilots flying at slow speeds and low altitudes. A WAC would not be satisfactory while a pilot is operating in the vicinity of Class B or Class C airspace. The charts are 20 × 60 in, which can be folded to the standard size of 5 × 10 in and have a scale of 1:1,000,000 (1 in equals 14 nm). WACs are identified by an alphanumeric group. Areas of coverage are contained in Fig. 3-7 for the contiguous United States, Mexico, and the Caribbean, and Fig. 3-8 for Alaska. (Note that the Alaskan CE-15 WAC covers the Canadian west coast down to Vancouver Island.) They are revised annually, except for a few in Alaskan (CC-8, CD-10, and CE-12) and Central American charts CH-22, CJ-26, and CJ-27) that are revised every 2 years. Features include:

- Visual aids to navigation
- Radio aids to navigation
- Airports
- Restricted areas
- Obstructions
- Topography

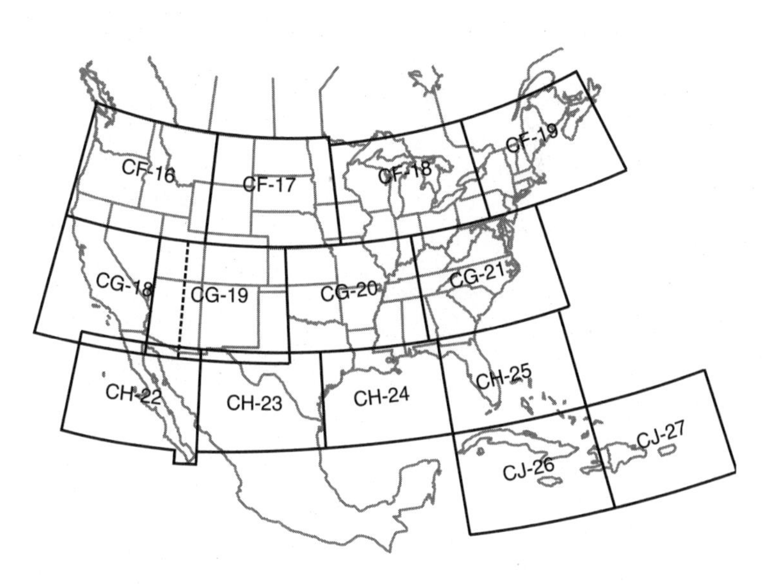

3-7 *World aeronautical charts are designed for visual navigation of moderate-speed aircraft.*

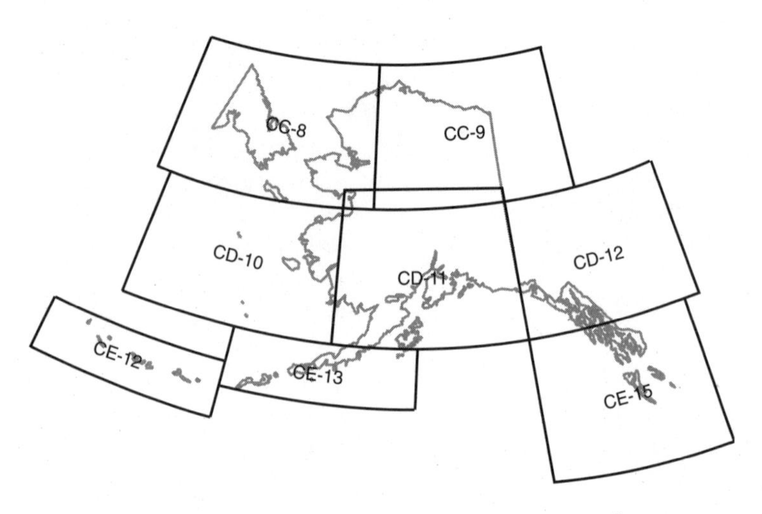

3-8 *World aeronautical charts are revised annually, except for a few in Alaska and Central America that are revised biennially.*

- Shaded relief
- Latitude and longitude lines
- Airways
- Other VFR-related data

I prefer WACs for cross-country flying. Refolding sectionals is cumbersome, even at the speeds of the Cessna 150, and WACs reduce cabin clutter. One WAC covers about the same area as four sectionals, shaving approximately one-fourth off the chart bill. These are the major advantages of the WAC. The biggest disadvantage is scale. You might recall my episode misidentifying the Ohio River, which occurred in 1971.

Sometimes we fail to learn from our mistakes. The following occurred in 1992.

Case Study

It was the second day of the trip back from the Oshkosh Fly-In. We had departed Gillette, Wyoming, for Pocatello, Idaho, in a Bonanza. Using a WAC chart, we planned to use VOR and pilotage navigation through Jackson Hole to Idaho Falls. The weather was not a factor, but visibility was restricted by smoke from numer-

ous forest fires. In this part of the country even at 12,500 ft, we were still below the peaks. After passing what I identified as the Grand Tetons we turned southwest toward Pocatello.

Well, you guessed it, we couldn't receive any VOR or establish communications with any facility. Continuing to fly down what I thought was the Snake River Valley, things didn't seem quit right. We followed an old aviation axiom: Follow a river or a road and it will normally bring you to a town, and hopefully an airport.

Even with 2 1/2 hours of fuel we decided it was time to resolve the issue of position. Since we were unable to establish communications on standard frequencies, I selected 121.5 MHz. We were not in distress, but there was a sense of urgency. Therefore, as outlined in the Aeronautical Information Manual, I broadcast "PAN PAN PAN" followed by the aircraft identification. Almost immediately a military Air Evac flight responded. Based on our assumed position, the Snake River Valley, Air Evac provided us with a frequency for Salt Lake Center.

After several tries, however, we were unable to establish communication. By this time we had come across a small town with a good-size airport. Unfortunately, there was no name on the airport. Since our transponder was being interrogated, we knew someone had us on radar. I selected 7700 and again broadcast "PAN PAN PAN." Air Evac again responded. I requested Air Evac to ask center to look for a 7700 squawk. In a few moments Air Evac responded with another center frequency.

Calling center, the controller immediately responded, "Your position is 6 miles east of Big Piney." That left us with one minor question: Where is Big Piney? After a few moments shuffling the chart we were on our way, although not by the route originally planned. As Maxwell Smart, Agent 86 of Control, would say, "Missed it by that much!" Well, it isn't much on a WAC chart.

I usually back up WACs with sectionals. Some IFR pilots routinely carry WACs in case of electrical failure or other emergencies where having a visual chart would be helpful.

Using visual charts

Consider consulting a flight planning chart for preliminary preparation of a long cross-country. My favorite is the jet navigation chart (JNC) series, discussed in Chap. 4. These charts help determine which WACs, sectionals, and TACs are required. Specific charts will depend on the route, possible alternatives, and the mission.

Ensure chart currency; if a chart is to be revised next month and the proposed flight is 6 weeks off, wait as long as possible for that new chart to finalize flight planning. Charts can be obtained from many sources; most pilot supply stores and FBOs carry charts, they are available through subscriptions from a number of sources, and they can be obtained from the government. Addresses and telephone numbers for government chart agencies are contained in Appendix B. Smaller FBOs normally carry only charts for their immediate vicinity.

Pilot supply stores might carry charts for the entire United States and most of North America. Pilots should become familiar with outlets in their area and determine which charts are readily available. Obtain all the charts that might be required. As in my case over West Virginia, it's very embarrassing to end up needing a chart and not having it. Pilots usually realize this in the air or at an airport with limited chart selection. It's always better to have too many charts than too few.

Bridging the gaps

Planning flights across chart boundaries can become a problem because sectionals and WACs are printed on both sides. Refer to Fig. 3-9. Plotting these routes is accomplished in the following manner.

There are approximately 2 minutes of latitude overlap between the north and south sides of each chart, greater in

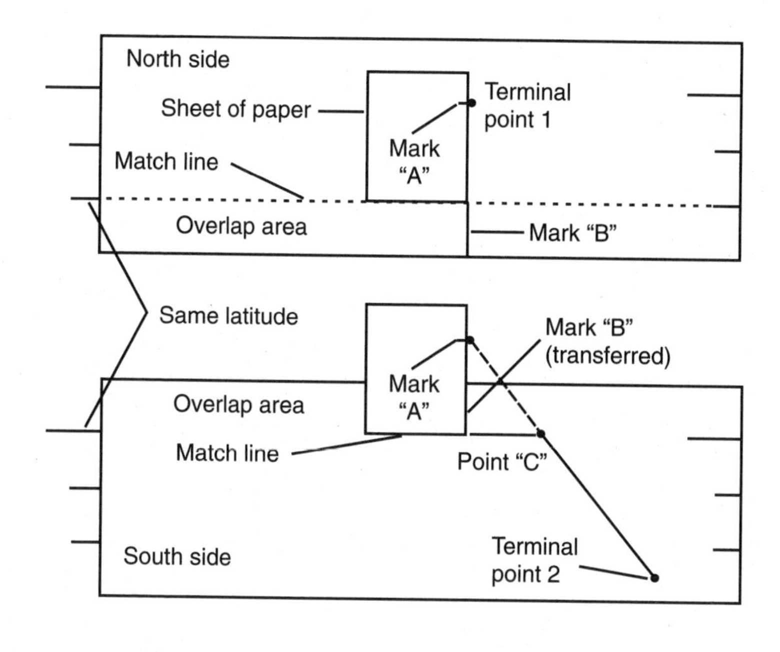

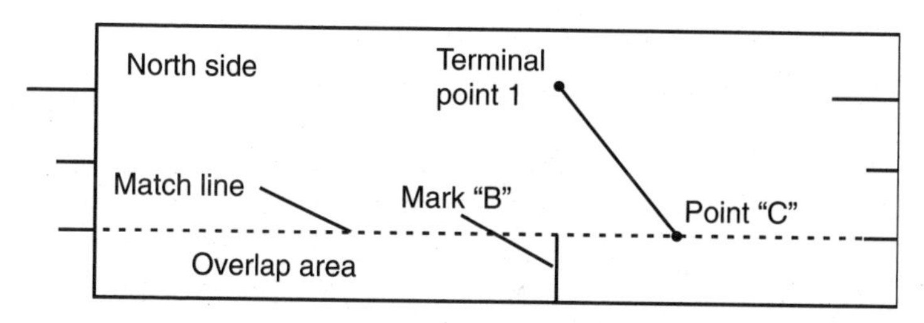

3-9 *Plotting course from the front to back side of charts can be cumbersome, but can be mastered with a little practice.*

certain cases. The pilot must first determine this overlap, and either visually note the position or draw two match lines that are common to both sides. To draw the match line on the north side, connect the latitude tick marks of the most southern minute of latitude and on the south side connect the marks of the most northern minute of latitude. These match lines must have the same latitude on north and south sides.

On the side of the chart having the terminal (departure or destination) nearest the match line, place a sheet of paper so that one edge corresponds to the match line and the other edge intersects the terminal airport. Mark the edge of the paper at terminal point 1, and label it "Mark A." Then make another mark on the chart extending from the match line to the edge of the chart, and label it "Mark B" as shown in Figure 3-9.

Turn the chart over and transfer Mark B to the other side of the chart, making sure to extend the mark to the match line. Align the sheet of paper to the match line with the corner of the sheet at the transferred Mark B. With a plotter, or other straightedge, align terminal point 2 with Mark A, and draw a line from the match line, now called Point C, to terminal point 2 as shown in Figure 3-9.

Turn the chart over again and transfer Point C to the other side of the chart. This can be done by measuring the distance from Mark B to Point C. With a plotter, draw a line from Point C to terminal point 1 as shown in Figure 3-9.

A direct course now consists of the line segment from terminal point 1 to Point C on one side of the chart, and from Point C to terminal point 2 on the other side of the chart. Be careful with this procedure. Errors occur from not properly considering the overlap area, incorrectly transferring Mark B or Point C, or including the overlap when measuring total distance. This procedure is a little complex, but can be mastered with a little practice.

Now we're ready to apply the principles and knowledge from Chap. 2, and the preceding portion of this chapter. Let's plan a pilotage flight from the Livermore Municipal Airport, Livermore, California, to the Reno Tahoe International Airport, Reno, Nevada. Review the WAC in Figure 3-10 for initial planning. For relatively short flights WACs might serve as excellent planning charts because of their scale and detail. For our flight there is an added advantage in that we can view the entire route on the WAC; on the sectional, departure and destination airports are on different sides of the chart.

The Livermore airport is located adjacent to the San Francisco Class B airspace. Terrain is mostly flat through the Sacramento Valley. We will be some distance from the Sacramento Class C airspace. Should we choose to verify our route with NAVAIDs, we see that the Sacramento, Hangtown, Squaw Valley, and Mustang VORs are along our course. If you haven't noticed or

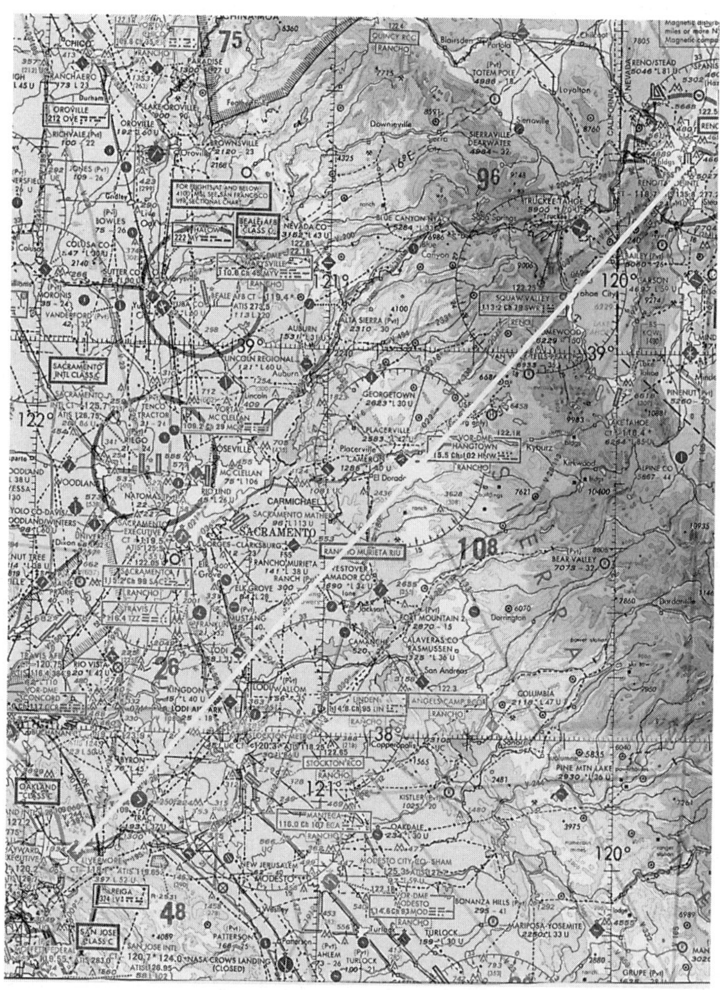

3-10 *WACs are often helpful with initial planning of short trips.*

are new to flying, you might not be aware that in the early days, NAVAIDs located in the vicinity of airports usually carried the same name as the associated airport. For example, the VOR just northeast of Reno was called Reno and the Squaw Valley VOR was called Lake Tahoe. To prevent any confusion or misunderstanding in our computerized ATC system, a program evolved to change the names of NAVAIDs that are not colocated with the airport of the same name. Policy dictates that names come from the immediate area. A diversion to ward off cross-country boredom might be reviewing airport VOR names and determining the reference; if the reference is not obvious, perhaps a hidden meaning can be explained by a pilot in the area—for example, *S*qua*W* Valley *R*esort (SWR).

For the portion of the route over the Sierra Nevada Mountains, maximum elevation figures (MEFs) indicate terrain from 9600 to 11,100 feet MSL. This might pose a problem for low-performance aircraft, especially considering weather and density altitude. Finally, we see that the destination is also located within an Class C airspace. We might want to consider the note just north of Reno that states, "Magnetic disturbance exists in the area extending 50 miles or more N.W., W, & S.W. of Reno, Nevada. Magnetic compass might not be accurate at low altitude." Apparently gold and silver are in those hills (?).

From this preliminary view we would also check for any SUA, such as restricted areas and MOAs, or any other high density areas. It is worth noting that MTRs do not appear on WACs. Notes on the TAC reveal that a San Francisco TAC is necessary for flight below 10,000 ft MSL, which is the top of the Class B airspace. The WAC refers us to the San Francisco sectional for flight below 4100 ft in the Sacramento area and below 8400 ft in the Reno area, the upper limits of Class C airspace. The WAC would be perfectly acceptable for a flight in good weather from Livermore to, for example, Carson City (south of Reno).

We could fly east of Livermore and follow the highway to Stockton. The Stockton airport has a control tower so we'll want to stay above their Class D airspace or obtain the required clearance. Proceeding northeast of Stockton toward the large reservoirs, we could pick up the highway and proceed south of the Lake Tahoe Airport, which also has a tower, then to Douglas County, and finally into Carson City. This would be unwise in poor weather because it is too easy to misidentify landmarks from the WAC, pick the wrong canyon, and end up boxed in; it has happened.

Let's move on to the sectional. A direct route has been established on the sectional chart, as illustrated in Fig. 3-11 (south portion) and Fig. 3-12 (north portion). A direct flight will take us out over the coastal mountains where the MEF is 4200 ft. We review the sectional, end panels, and margin.

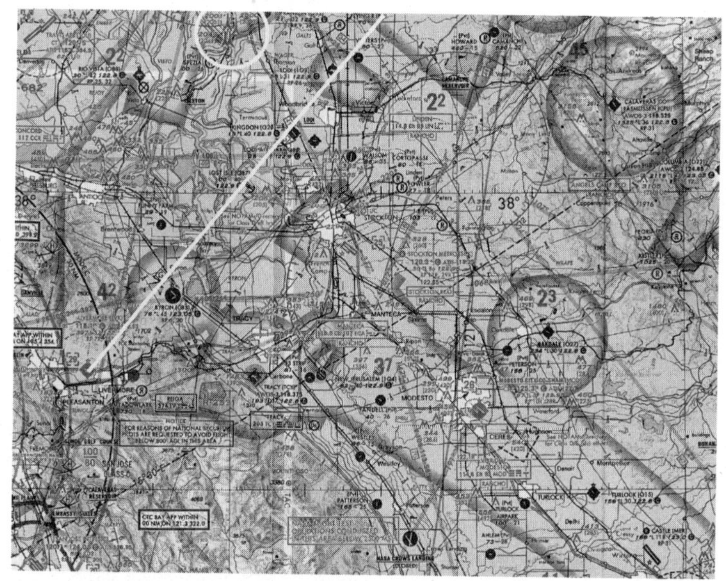

3-11 *The sectional is satisfactory for operating outside of the Class B airspace, but operations close to or under the Class B require the appropriate terminal area chart.*

The sectional provides enough information to negotiate Class C airspace. However, operating in or around Class B airspace requires the use of the appropriate TAC.

Once clear of the Livermore Class D airspace, we see our course will take us out over the Sacramento River Delta. Just southwest of Franklin are a number of towers that, from the symbols, extend above 1000 ft. From the chart, these towers extend to 2001 ft MSL. These towers are equipped with strobe lights that help but are still hard to see, especially in haze or fog. Realize that large towers usually have guy wires for support, which can easily snag an aircraft flown by an unsuspecting pilot. Flying at low altitudes in low visibility—even while complying with all the regulations—makes these feature hazardous.

Terrain starts to rise over the Sierra foothills. Critical elevations range generally between 8000 and 11,000 feet. Donner Pass, just west of Truckee, has an elevation of 7088 ft. The Spooner Summit Pass, on the east central side of Lake Tahoe, has an elevation of 7146 ft. East of the passes it's

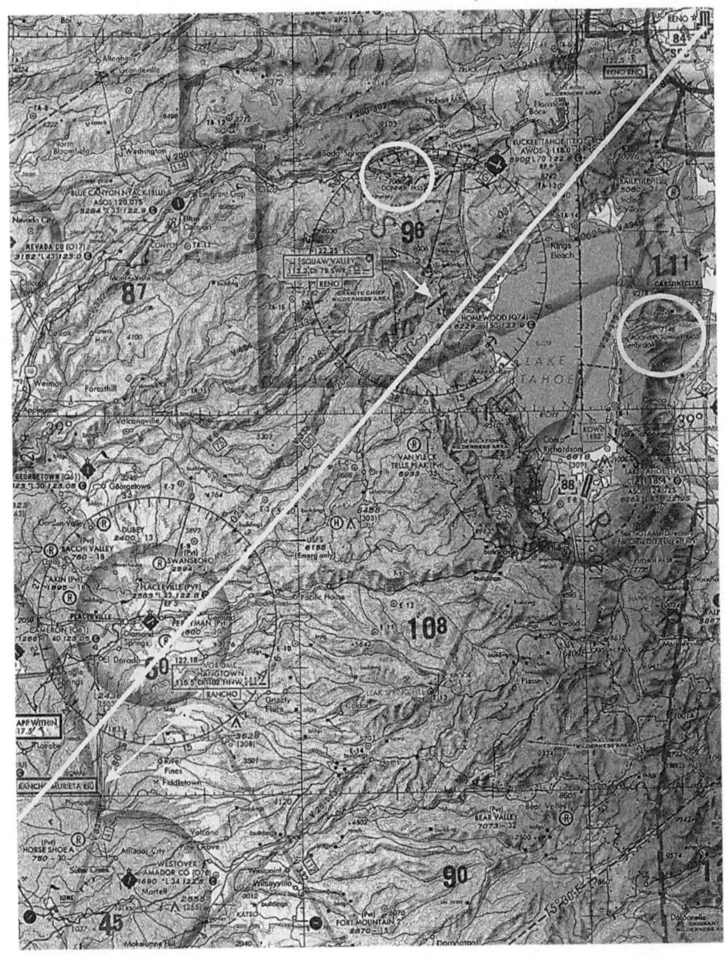

3-12 *A sectional is required for operating into or in the vicinity of Class C airspace.*

generally all downhill through the valleys to Reno, which has a field elevation of 4412 ft.

Notice what appears to be a long runway on the west side of Lake Tahoe, just south of the Squaw Valley VOR. A gotcha flight instructor-to-student question is what does this symbol represent? The answer is at the end of the chapter.

Reno's Class C airspace extends from the surface to 8400 ft MSL. From the chart or end panel, we determine the approach frequency is 119.2 MHz. The chart also provides ATIS (135.8) and tower (118.7) frequencies; ground frequency for this airport is 121.9. The only other frequency we might need is the Reno FSS, which the chart indicates is 122.5.

From our review of the sectional we determine there are no military training routes along our proposed course. If there were, we could note the route numbers: four-digit numbers for routes flown at or below 1500 ft and three-digit numbers for routes flown above 1500 ft AGL. If our flight altitudes are above 1500 feet AGL we can disregard any routes with four digits. Call an FSS for operational details along any conflicting routes.

Now for the flight from Livermore to Carson City using the sectional, weather is marginal, ceilings are between 1000 and 3000 ft, and visibility is 3 to 5 miles. We'll definitely need the sectional to establish the limits of controlled airspace. We will observe all regulations as they apply to controlled airspace, distances from clouds, and minimum safe altitudes. Departing east of Livermore we could cruise at 1500 ft MSL, and even lower after crossing into the valley and through the Stockton area. Livermore has Class D airspace. Class E airspace based at 700 ft exists to the east, and along the highway to Stockton Class E airspace starts at 1200 ft AGL; therefore, at our altitude communication is required to transit Stockton's Class D airspace.

We could then proceed northeast of Stockton toward the Comanche and Hogan Reservoirs, then follow the highway through Carson Pass, which has an elevation of 8650 ft. The base of Class E airspace through this area is 1200 ft AGL, as indicated by a blue vignette. Where controlled airspace begins at other than this altitude, the chart is labeled in feet above mean sea level. For example, south of Minden, Nevada, the floor of Class E airspace is 12,300 MSL. From the Carson Pass we could then follow the highway into Carson City.

(This is not a recommendation to fly at low altitudes through mountainous areas. FAA: "The decision as to whether a flight can be conducted safety rests solely with the pilot." This depends on a pilot's training and experience.)

Could we fly from Livermore to Stockton with ceilings of 500 to 1000 ft and visibilities ranging from 1 to less than 3 miles?

Technically, yes. This flight would require a special VFR clearance out of the Livermore Class D airspace and into the Stockton Class D airspace. Enroute we would be required to remain clear of clouds below 700 ft AGL floor of Class E airspace east of Livermore, and then below 1200 ft to the boundary of the Stockton Class D airspace. Could this flight be conducted at night under the same conditions? No. At night, even in uncontrolled airspace, 3 miles visibility and standard distance from clouds is required. When you are flying with reduced ceilings and low visibilities, it's often a good a idea to slow down.

Case Study

> *Our destination was Crescent City, California. The VOR was out of service, which precluded an instrument approach. The coastal stratus tops were about 1500 ft, hugging the hills. Bases were reported at 600 ft with a visibility of 2 miles. We found a small hole in the stratus and obtained a special VFR clearance. To allow more time to see and avoid obstacles, or anything else, we slowed the Mooney to 100 knots. There is no rational reason to be flying in marginal weather at 160 knots.*

When flying in marginal weather, .be very careful not to impose on someone's airspace, especially if a preplanned route is altered. (The 1991 Hayward–Bakersfield–Las Vegas Air Race began with overcast ceilings between 1500 and 2500 ft AGL. The first checkpoint was the Pine Mountain Lake airport in the Sierra foothills at an elevation of 2900 ft. Most of us ended up bypassing this checkpoint because of the weather. Using pilotage, our crew flew to our next plotted checkpoint south of Pine Mountain to resume our preplanned course. Many pilots, after abandoning Pine Mountain, headed straight for the next checkpoint, apparently neglecting to consider or communicate with the folks controlling, at this time, the Castle AFB Class C airspace.)

San Francisco's terminal area chart will be used to navigate the first part of the flight, as shown in Fig. 3-13. Livermore airport is just outside of Class B airspace. Therefore, Class B airspace for our planned route of flight is not a factor. There is a NOTICE east of Livermore. It states: "FOR REASONS OF NATIONAL

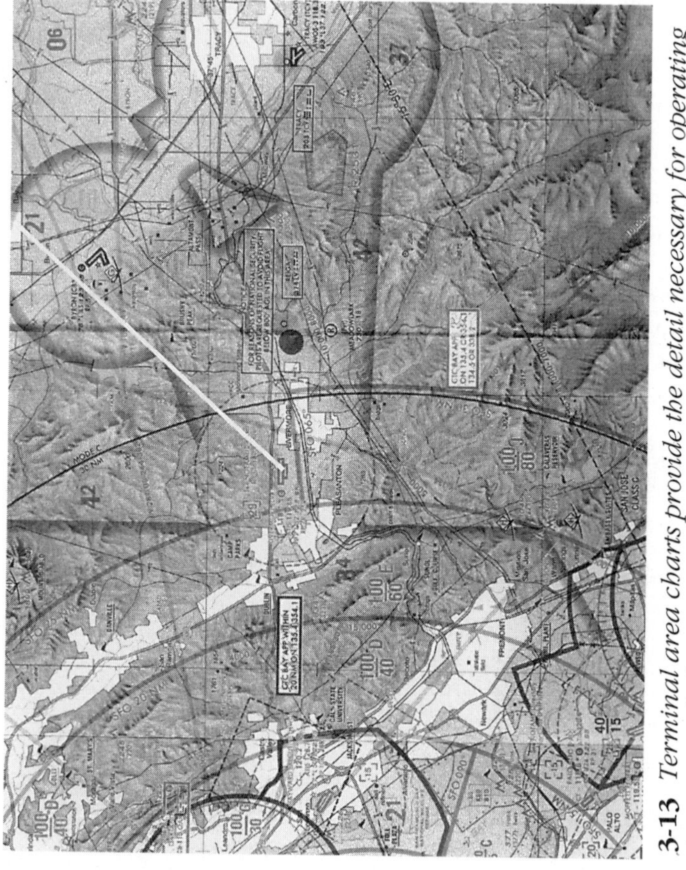

3-13 *Terminal area charts provide the detail necessary for operating in congested Class B terminal airspace.*

SECURITY PILOTS ARE REQUESTED TO AVOID FLIGHT BELOW 800′ AGL IN THIS AREA." Certainly fixed wing, at least, has no business operating below 1000 AGL in this area.

Take advantage of the detail available on the TACs, where they're available, especially in marginal weather. This chart would be useful for our Livermore to Carson City trip. From the chart, east of Livermore we see more clearly the transmission lines that cross the course. We can use these along with the railroad to update progress of the flight. The TAC shows that one of the railroad tracks goes through a tunnel on its second crossing of the highway, which is not seen on the sectional chart. Contours indicate that the approximate elevation of the pass between the Livermore Valley and the San Joaquin Valley is above 1000 ft, but lower than 1500 ft.

I obtained a DUAT weather briefing in the morning prior to departure and filed a VFR flight plan through the service. After clearing Livermore's Class D airspace, we switched to Oakland radio on 122.5 MHz to open the flight plan. We were south of Mt. Diablo, correcting the heading toward the planned flight course, and contacted Oakland flight watch on 122.0 for a weather update over the Sierras.

Checkpoint considerations

A primary checkpoint consists of a topographical feature, or set of features, that cannot be mistaken for any other place in the same general area. Often three or more secondary features can be combined to form a primary checkpoint.

Secondary checkpoints are small towns, streams, a single road or railroad, mountain range, or any other feature that could be mistaken for similar features in the same general area.

A primary checkpoint could be a single feature, such as Arizona's Meteor Crater. It is large and unique, and cannot be mistaken for any other feature in the same area. Several smaller features can be combined to form a primary checkpoint. Unless the features are relatively distinctive, such as an airport and adjacent town, three features should be used: a town, highway, and railroad. One feature, or even two, can be mistaken, and should if at all possible be avoided.

Pilots should consider the availability of suitable checkpoints during flight planning. This is the time to select an alternate route if you're not comfortable with what's available. If you do get lost, or even think you're lost, call for assistance before a relatively simple flight assist becomes an accident.

In addition to a suitable checkpoint, pilots should consider the type of terrain and possible emergency landing sites. Hazardous terrain can consist of mountains, deserts, and large bodies of water. High, rugged mountains may afford few, if any, suitable emergency landing sites. Take for example the following situation.

Case Study

> *We had planned a flight from Fresno, California, to participate in an air show at the Mammoth Lakes airport east of the Sierra Nevada Mountains. We had a turbocharged Cessna 210 for the flight. Since I had not flown the route through the Kearsarge Pass, elevation 13,291 ft, I thought this would be a good opportunity. The weather was perfect. East of Fresno we climbed to 13,500 ft. This was within a few hundred feet of pass elevation and still well below the mountain peaks! The steep mountain slopes and narrow mountain valleys left very few emergency landing sites. I have since decided this is not an acceptable risk, even in perfect weather.*

Deserts, especially in the United States southwest, may offer little, if any, assistance to a downed aircraft and its crew. Pilots should carefully consider flight over these sparsely populated areas. (Recall my first experience with a swamp described in Chap. 2.) Usually only a deviation of a few miles will take the pilot over a highway and settlements that can offer assistance in case of an emergency.

In an emergency, flight over large bodies of water could put the pilot in a situation that requires ditching the aircraft. What survival equipment do we have on board? Have we ever practiced or even considered a ditching maneuver? If we survive the landing, how long can we expect to survive in the water? These are all questions that should be considered. Many pilots in southern California routinely fly to Catalina

Island, 20 miles off the coast. When I fly there, I always fly at 6500 ft over and 7500 ft back. In the case of an engine failure, I have enough altitude to glide to shore.

A snow-covered landscape can produce additional hazards. Topography on the chart may be difficult to recognize under snow-covered terrain. Emergency landing sites may be far and few between. Major highways are cleared first, secondary roads and smaller airports may remain snow covered for days after a major snow storm.. Certainly flying over major, heavily traveled roads would reduce the risk of such operations.

Back on course

Returning to our flight, we see from the TAC that the Bryon airport has both parachute and glider operations. A good practice would be to monitor the CTAF or contact the controlling agency for traffic advisories in this area. Figure 3-14 shows the Byron airport, the major lake adjacent to the field, and the irrigation canals emanating from the lake. This makes an excellent checkpoint.

When well clear of Bay Area airspace and traffic congestion, switch from the TAC to the sectional chart for navigation. As we pass into the northern San Joaquin valley, we see a view of

3-14 *A good practice when operating in the vicinity of airports with parachute jumping and glider activity would be to monitor the CTAF or contact the controlling agency for traffic advisories.*

the city of Stockton off to the southeast (Fig. 3-15). This area offers good primary checkpoints because of the relation of water, towns, and roads. A word of caution: be very careful flying in this area at low altitudes, especially in low visibilities. The numerous obstruction symbols on the chart represent many power lines that are stretched across the water.

Figure 3-16 shows a view of Stockton with its interstate highway. This city, with the highway, makes a good checkpoint. Remember, if at all possible stay away from sparse

3-15 *Good primary checkpoints are a combination of water, towns, and roads.*

3-16 *A major city with a prominent highway and secondary roads makes a good checkpoint.*

checkpoints, such as a small town and a road. Almost every small town has a road running through and especially in flat country many of the roads are parallel. Try to select checkpoints with three or more features.

An FSS communication box just north of our course in Fig. 3-12 reads RANCHO MURIETA RIU. Routine communications, such as position reporting, flight plan updating, or other FSS services are possible on the standard FSS frequency 122.2. Also notice the Hangtown VOR-DME data box. Rancho radio has a receive-only at that site on 122.1. We could contact Rancho radio by transmitting on 122.1 and listening on the Hangtown VOR 115.5; always advise the FSS which frequency is being monitored.

As we pass Placerville and the Hangtown VOR, we see Highway 50, a major secondary road as it winds its way into the Sierra foothills. This is shown in Fig. 3-17. Such landmarks often make good progress checks. A disadvantage of this type of checkpoint is that it may be difficult to determine if we are off course—left or right, and how far.

Approaching the Sierra foothills we begin a climb to 11,500 ft. Off to our left is Folsom Lake, shown in Fig. 3-18. These lakes often make excellent checkpoints; the shape of the lake and the dam are easily verifiable with the chart. Compare the presentation of these lakes on the WAC in Fig. 3-10, with the lakes on the sectional charts in Figs. 3-11 and 3-12.

The chart shows many small lakes along the Sierra foothills. An aerial view is illustrated in Fig. 3-19. Care must be exercised selecting them as checkpoints; they are numerous and their shapes are similar. Take, for example, the following incident.

Case Study

We had attended the Reno Air Races in 1970 and were returning to Van Nuys via Fresno, California. I had flight-planned down through Minden, Nevada, over the Sierra Nevada Mountains into Fresno. For radio navigation all we had was a VOR receiver. Crossing the Sierras we were confronted with a scene similar to that in Fig. 3-19. Unable to make heads or tails of our position,

3-17 *Secondary roads often make good progress checks. A disadvantage is that it may be difficult to determine position left or right of course, and how far.*

> *I called the then Fresno FSS for a DF steer. Shortly after we had our position located and proceeded uneventfully to Fresno.*

Approaching the crest of the Sierras, the terrain rises rapidly and the aircraft is buffeted by some turbulence. The lowest terrain is south of a direct course, over the center of Lake Tahoe. As we approach the lake we are greeted by the scene in Figure 3-20. We can tell that we will clear the crest of the mountains because the terrain beyond appears to be descending in relation to the crest of the mountains. (We were over

3-18 *Often the shapes of lakes and their dams make easily verifiable checkpoints.*

3-19 *Care should be taken in the use of lakes as checkpoints when there are many similar lakes in the same general area.*

the lake and decided to continue through Spooner Summit Pass, then over Carson City, and into Reno.)

East of Lake Tahoe we again contact flight watch and file a pilot report regarding conditions over the mountains. Frequent, objective pilot reports cannot be overemphasized, even if the weather and the ride are clear and smooth. Reporting enroute is a timely practice that makes current

3-20 *We can tell we're above the crest of the mountains because the terrain behind appears to be descending in relation to the crest.*

information available to briefers and forecasters, and ultimately other pilots.

After crossing the mountains we can see Reno off in the distance, as shown in Fig. 3-21. The VFR flight plan was closed over Carson City via Reno radio prior to entering Class C airspace. The ATIS report is noted prior to contacting approach control for entry into the Reno Class C airspace. Recall that frequencies are already listed; therefore, we concentrate on flying the airplane, looking for traffic, and navigating, rather than fumbling with the chart trying to find a frequency. We pass Steamboat and follow the highway to the Biggest Little City in the West.

Coastal cruise

The value of a TAC in congested terminal airspace cannot be overemphasized. Let's take a flight from Livermore, in one of the coastal valleys, to Half Moon Bay, on the coast, southwest of San Francisco. A portion of the San Francisco TAC is contained in Fig. 3-22. A direct flight will take us through the Oakland Class C and the San Francisco Class B airspace. Departing Livermore we will probably wish to stay below 4000 ft because of aircraft inbound to Oakland and Hayward

3-21 *With our destination in sight, we may wish to close our flight plan in the air, before switching over to ATIS, approach control, and the tower.*

at that altitude. This information is obtained from the VFR flyway planning chart on the reverse side of the TCA, and illustrated in Fig. 3-23.

The Class C airspace in the vicinity of Hayward extends from 1500 ft MSL to the base of the Class B airspace. We can either skirt south of the Oakland Class C airspace, or contact approach on 135.4 for Class C airspace services through the area. Don't forget about the Hayward Class D airspace. The San Francisco Class B airspace over Hayward extends from 2100 ft to the top of the Class B airspace 10,000 feet; then from the vicinity of San Carlos from 1500 ft. We will need to avoid the surface Class B over San Mateo or obtain a Class B clearance. Otherwise, crossing the bay we will need to obtain clearance through either the San Carlos or Palo Alto Class D airspace. This is often easier than trying to obtain clearance through the Class B airspace itself. Tower controllers might assign aircraft specific routes to avoid their airport traffic patterns.

Beyond the San Carlos Class D airspace, the base of the Class B airspace is 4000 ft. Now we can climb safely over the hills and proceed into Half Moon Bay. We need to be very careful of minimum safe altitudes and required cloud

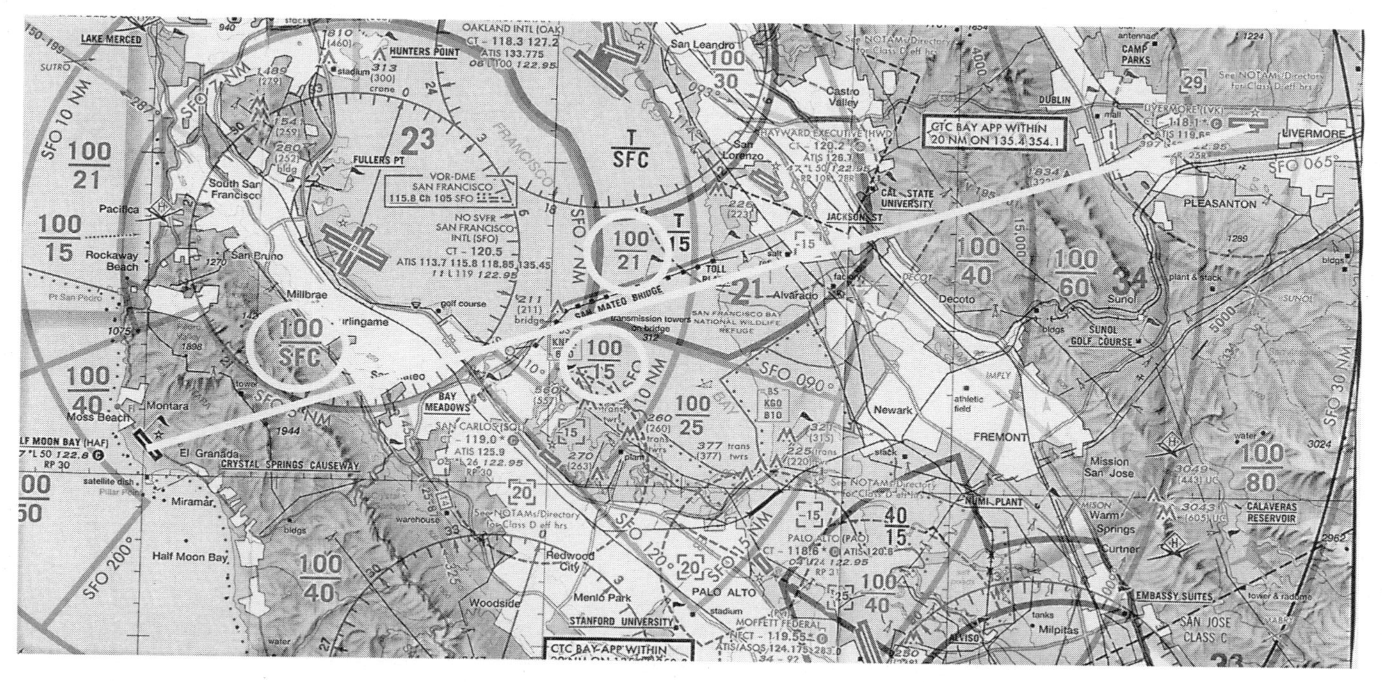

3-22 *The value of a TAC cannot be overemphasized.*

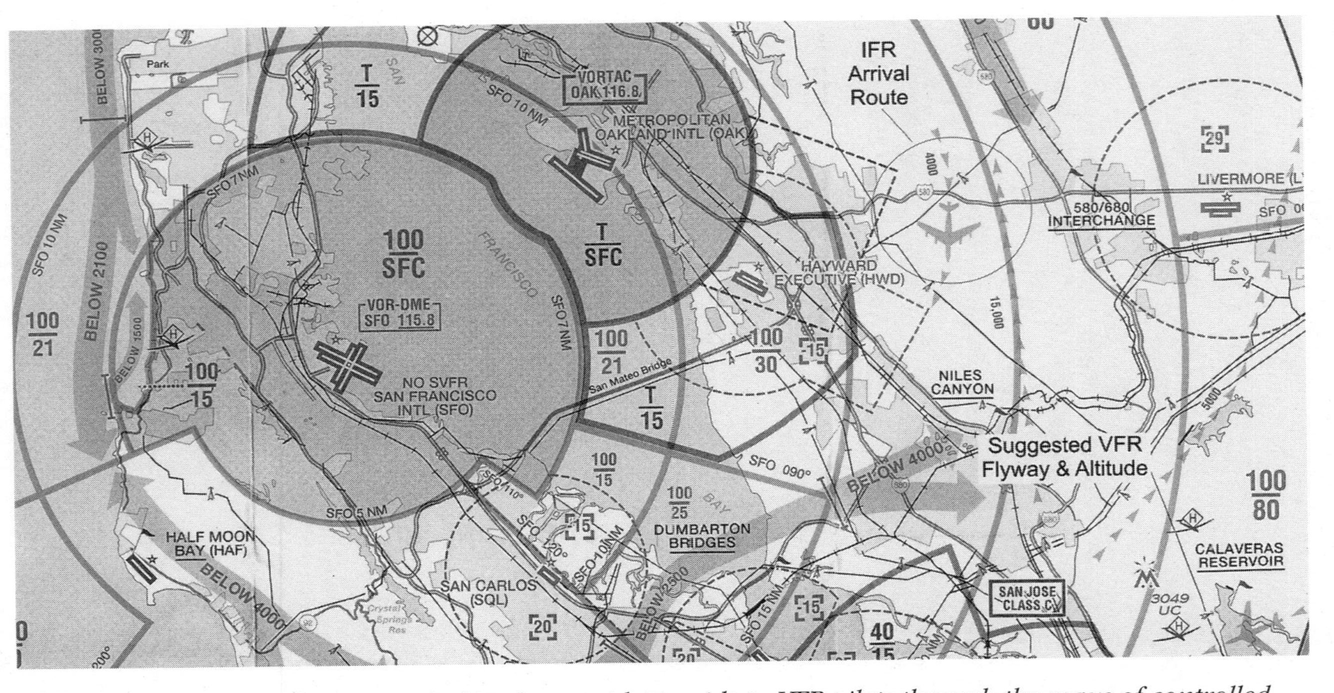

3-23 *"Suggested" VFR flyways and altitudes provide a guide to VFR pilots through the maze of controlled airspace; they do not necessarily guarantee a safe flight.*

clearance and visibility, and keep a sharp eye out for other traffic. Pilots of high-performance aircraft might wish to slow down in congested airspace in the vicinity of a Class C or Class B airspace.

An alternative would be to request clearance through San Francisco's Class B airspace. This would certainly be an option. However, like everything else there are limitations. You might recall my experiences with Bay Area controllers on just such a flight.

Notice that the VFR flyway chart in Fig. 3-23 shows "suggested" VFR flyways and altitudes. These routes are provided to guide VFR pilots through the maze of controlled airspace. They, however, do not necessarily guarantee a safe flight. For example, the base of Class B and Class C airspace often lies just above these altitudes. This is illustrated in Fig. 3-23 in the vicinity of the Dumbarton Bridges. Pilots must be prepared for, possibly severe, wake turbulence encounters in the areas.

A bit earlier the reader was asked about the symbol adjacent to the Squaw Valley VOR on the west side of Lake Tahoe. If you look northwest to southeast, from just above the Squaw Valley NAVAID box, to just south of the Lake Tahoe Airport, then south-southwest of the Alpine County Airport, you'll see the large letters: S-I-E-R-R-A. The so-called runway is the I in SIERRA. You might want to have some fun and ask your fellow pilots, or better yet an instructor, to decode this symbol.

4

Supplemental charts

The Coast and Geodetic Survey introduced a new family of aeronautical charts in the early 1950s to simplify high-speed jet and transport navigation as well as small aircraft visual flying. The series included planning, radio facility, approach and landing, as well as visual charts. These included jet navigation charts (JNCs) for visual flying and new experimental approach and landing charts for instrument operations.

On April 4, 1991, the Grand Canyon visual flight rules (VFR) aeronautical chart became available. The chart depicts communication and minimum altitudes for flight over the canyon in accordance with Special Federal Aviation Regulations.

Pilots have access to a number of supplemental visual charts through the National Aeronautical Chart Office (NACO) and other sources, notably Canadian charts published by the Canada Map Office, Department of Natural Resources. NACO maintains a public sales program and publishes a catalog that contains descriptions, availability, prices, and ordering procedures for National Imagery and Mapping Agency (NIMA) aeronautical products, primarily covering foreign regions.

Canada produces charts similar to their United States counterparts: WACs, VFR navigation charts (sectionals), and VFR terminal area charts. A Canadian chart catalog and products are available by mail or through various authorized dealers located at airports and cities throughout North America. Canada Map Office address, telephone numbers, e-mail, and Internet site are contained in App. B.

National Imagery and Mapping Agency (NIMA) visual charts

The National Imagery and Mapping Agency produces a series of visual navigation charts, mainly in support of military missions. Some maps may be adapted for civil flight planning and navigation. Basic NIMA charts consist of global navigation and planning charts, LORAN charts, operational navigation charts, tactical pilotage charts, and joint operations graphic charts.

Global navigational and planning charts

Global navigational and planning charts (GNC) are designed for flight planning, operations over long distances, and en route navigation in long-range, high-altitude, high-speed aircraft. GNC scale is 1:5,000,000 (1 in equals 69 nm). Sheet size is approximately 42 × 58 in. Polar regions use the transverse Mercator projection, and other regions, the Lambert conformal conic projection.

The global navigation chart series serves as the base for production of loran navigation charts and aerospace planning charts (ASCs). Features include:

- Principal cities
- Towns
- Drainage
- Primary roads
- Primary railroads
- Prominent culture
- Shaded relief
- Spot elevations
- NAVAIDs
- Airports
- Restricted areas

Figure 4-1 contains an excerpt from a global navigational chart. Shaded relief contains tints indicating relatively flat areas and those with steep relief, along with spot and critical elevations; however, contours and gradient tints are not included. Cities of strategic or economic importance, major

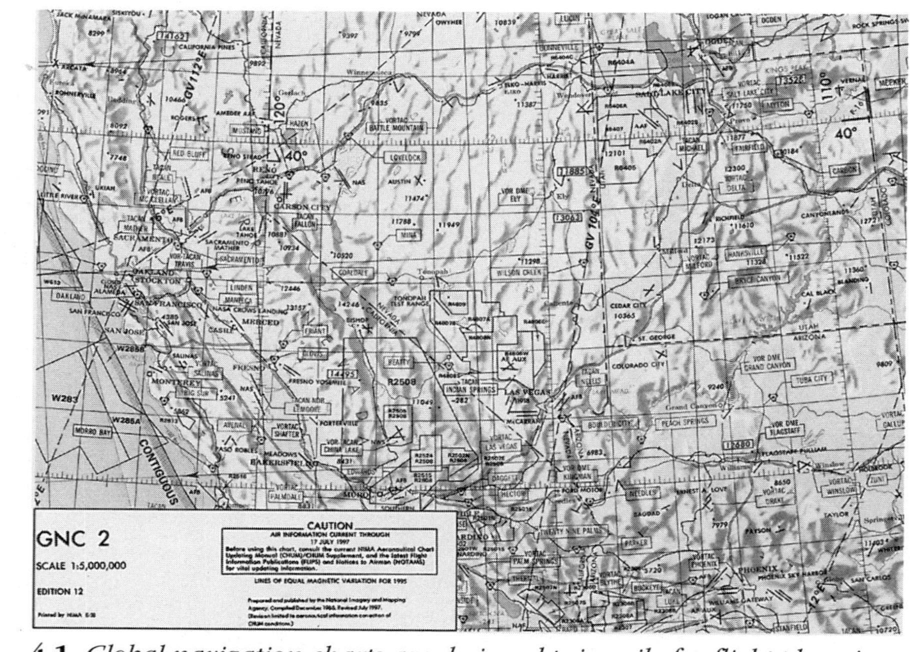

4-1 *Global navigation charts are designed primarily for flight planning, operations over long distances, and en route navigation in long-range, high-speed, high-altitude aircraft.*

towns, primary road and railroad networks, and other significant cultural features are displayed. Hydrography includes open water vignette, coastlines, and major lakes and rivers.

In Fig. 4-1, a caution note warns, "Before using this chart, consult the current NIMA Aeronautical Chart Updating Manual (CHUM)/CHUM Supplement, and the latest Flight Information Publications (FLIPS) and Notices to Airmen (NOTAMS) for vital updating information." The CHUM and FLIP are in effect the military version of the *Airport/Facility Directory* and are discussed in chapter 9.

These charts can be used for wall display because one sheet covers the United States, Canada, and part of Alaska. They are suitable as planning charts because of their relief and major cultural features, for example, for plotting a flight from San Francisco to Salt Lake City. From the chart we can see that a direct route would be over rough, sparsely populated terrain. The lower terrain would be through Reno, Battle Mountain, and Elko, then to Salt Lake City. This was the original airmail route. An alternative, although much longer route would be through central California to Las Vegas, Cedar City, and then to Salt Lake City. This route would generally be over lower terrain, perhaps a preferable route in the winter months.

Loran charts

Loran charts provide a plotting area where ground wave and sky wave correction values have been printed for loran navigation. Loran lines on these charts furnish a constant time difference between signals from a master and slave Loran station; however, with most of today's units, which incorporate microprocessors, Loran units provide the pilot with direct position readout, along with course, speed, and distance to specified locations. Chart scale is 1:3,000,000, sheets are approximately 42 × 58 in.

With the implementation of GPS, continued funding of loran is in doubt. At present there are only three polar charts available. The only chart that will continue in production covers northeast Canada and Greenland. The charts include spot elevations, solid land tint, major cities, coastlines, and major lakes and rivers.

Jet navigation charts

Jet navigation charts (JNCs) are suitable for long-range, high-altitude, high-speed navigation. Chart scale is 1:2,000,000. Features include:

- Cities
- Major roads
- Railroads
- Drainage
- Contours
- Spot elevations
- Gradient tints
- Restricted areas
- NAVAIDs
- Broadcast stations
- Airports
- Runway patterns

Runway patterns are exaggerated so they can be more readily identified as visual landmarks.

JNCs are available for the world; three charts cover the United States. The charts that cover the United States can be combined into a reasonably sized wall map. These charts are ideal for planning purposes because they have better terrain information, which is important to VFR pilots flying relatively low-performance aircraft.

Operational navigation charts

Operational navigation charts (ONCs) support high-speed radar navigation requirements at medium altitudes. Other uses include visual, celestial, and radio navigation. These charts have a scale of 1:1,000,000, the same as WACs. ONCs are available for all of the land masses of the world. Sheet size is approximately 42 × 58 in, covering 8° of latitude.

Figure 4-2 is an example of an ONC that covers the same general area as the WAC excerpt in Fig. 3-10. Notice that ONCs do not provide communications, airways, or airspace information, except for restricted, military operations, and

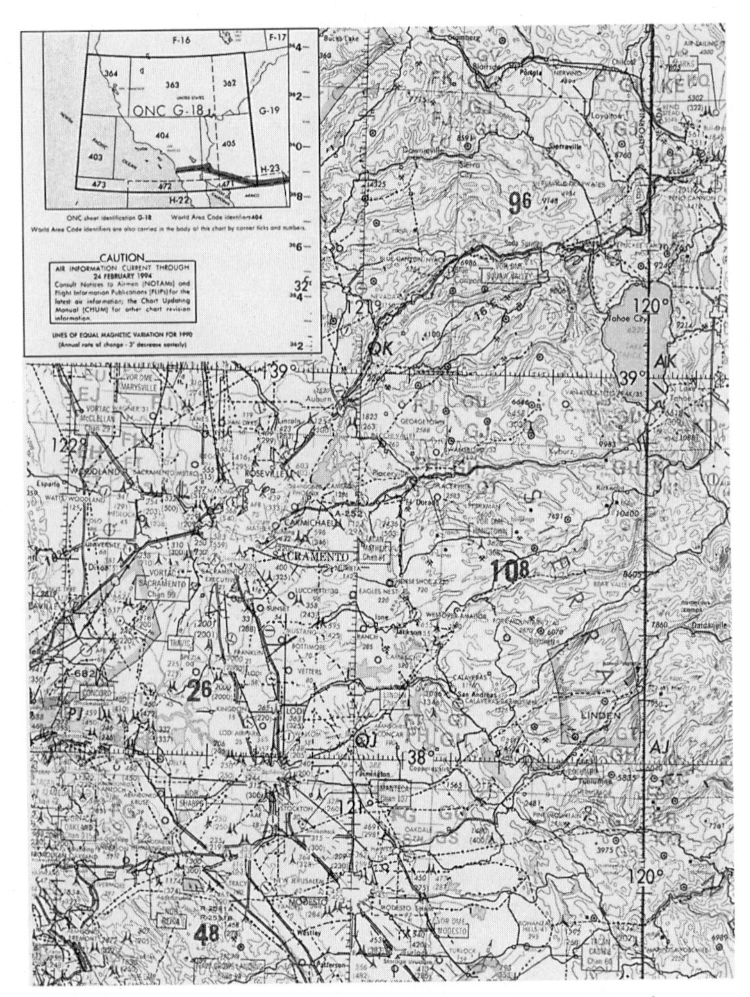

4-2 *Operational navigation charts support high-speed navigation requirements at medium altitudes.*

alert areas. Navigational information is current only through the date stated on the chart. Pilots are advised to consult NOTAMs and FLIPs for the latest information and the CHUM for other chart revision information.

Operational navigation charts are identified in the same manner as WACs. Starting at the North Pole with the letter A, each successive row of charts uses the next letter of the alphabet through X, which covers the South Pole. Each row of charts is labeled with a number that generally begins at the prime meridian, with subsequent numbers to the east. Because only land masses are charted in this series, chart G-18 will not necessarily be located under chart F-18. In Fig. 4-2, the chart joining G-18 to the north is F-16.

Because of the lack of aeronautical information, ONCs are not suitable for flights in the United States; however, these charts might be useful for pilots planning to fly in other countries where WACs are not available.

Tactical pilotage charts

Tactical pilotage charts (TPCs) support high-speed, low-altitude, radar, and visual navigation of high-performance tactical and reconnaissance aircraft at very low through medium altitudes. Tactical pilotage charts cover one-fourth the area of operational navigational charts. They are identified by the respective ONC letter and number and an additional letter representing the TPC (TPC G-18A). TPCs are not available for all areas of ONC coverage. TPCs have a scale of 1:500,000, the same as sectional charts.

Figure 4-3 contains an example of a TPC covering the same general area as the sectional in Fig. 3-11. Like ONCs, tactical pilotage charts do not provide communications, airways and fixes, or controlled airspace; airports, shaded relief, and topography are similar to the sectional.

Tactical pilotage charts are not suitable for navigation in the United States because of the lack of aeronautical information; however, these charts might be useful for pilots planning to fly in other countries. Some pilots like to obtain these charts when planning trips outside the United States just to get the lay of the land.

Joint operations graphics

Joint operations graphics—air (JOG-A) are suitable for preflight and operational functions. Scale is 1:250,000, the same as TACs. Figure 4-4 contains an example of a JOG-A, which covers the same general area as the TAC in Fig. 3-13. Communications, airways and fixes, and controlled airspace are not indicated.

4-3 *Tactical pilotage charts are designed for high-speed, low-altitude radar and visual navigation.*

JOGs are not available for sale outside of the United States. These charts could serve helicopter or other operators where low-level navigation is required and NACO TACs are not available; however, they would have to be used in conjunction with the associated sectional for proper communications and to establish the limits of controlled airspace.

Military training route charts

The FAA issued a waiver to the Department of Defense (DOD) in 1967 to conduct various training activities below 10,000 ft MSL at speeds in excess of 250 knots. These activi-

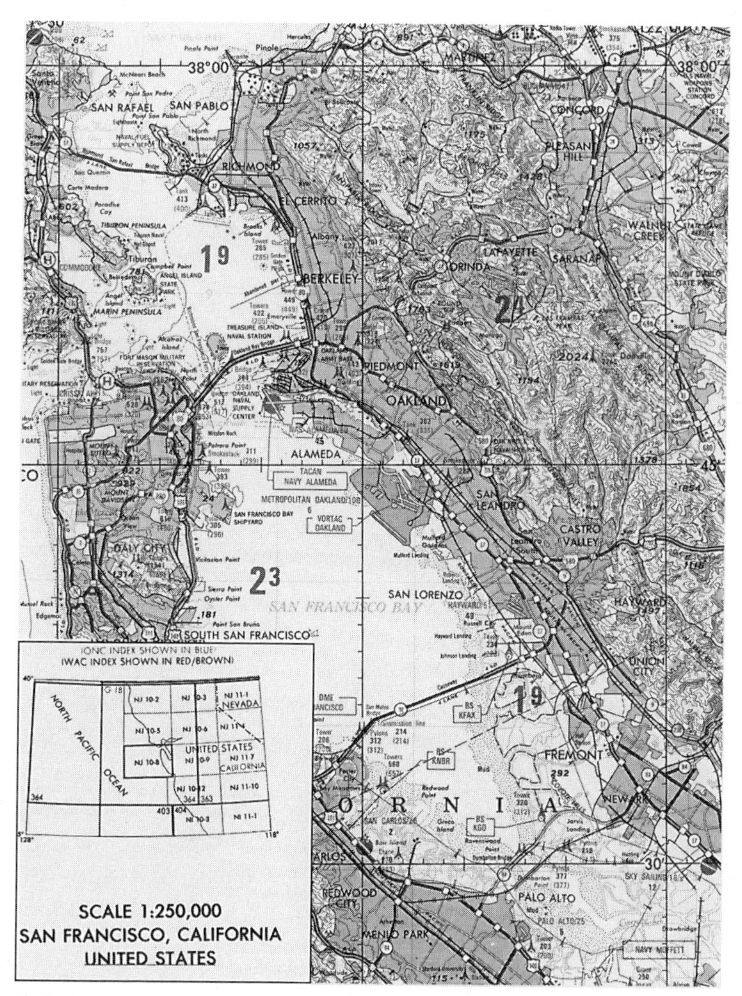

4-4 *Joint operations graphics are suitable for preflight and operational functions.*

ties included low-altitude navigation, tactical bombing, aircraft intercepts, air-to-air combat, ground troop support, and other operations in the interest of national defense. The number and complexity of these routes were to be limited to that considered absolutely necessary. Route widths vary from 2 to 10 nm. En route altitudes will be the minimum necessary for operational requirements but in no case at altitudes less than those specified in the regulations for minimum safe altitudes. They range from 500 ft, or lower, to higher than 10,000 ft. Active times vary and are specified for each route, ranging from daylight hours, Monday through Friday, to continuous. Routes are designed to be clear of Class B, Class C, and Class D airspace. Additionally, to the extent possible, routes remain clear of populated areas, Class E airspace, and uncontrolled airports.

Fact

Used to support the former Strategic Air Command (SAC) low-level missions, these routes were called "oil burners." This was in reference to the smoky jet engines of the time. Mysteriously, after the oil crisis of the early 1970s these routes were changed to "olive branches." Go figure.

Military training routes fall into two categories: IFR military training routes (IRs) and VFR military training routes (VRs). VRs are established only when an IR route cannot accommodate the mission. IRs might be flown in all weather conditions. VFR routes are flown only when forecast and encountered weather conditions equal or exceed 5 miles visibility and a 3000-ft ceiling. Recall from Chap. 2 that IRs and VRs operated at or below 1500 ft AGL will be identified by four-digit numbers (IR 1007, VR 1009). Operations conducted above 1500 ft AGL are identified by three-digit numbers (IR 205, VR 257). Route widths vary for each MTR and can extend several miles on either side of the charted centerline.

The National Imagery and Mapping Agency (NIMA) publishes military training route (MTR) charts. The charts provide a visual depiction of routes, along with a specific route number. Three charts are published for the United States: western, central, and eastern. Charts are published every 56 days and are available by single copy or annual subscription. The Department of Defense (DOD) provides these publications to flight service stations for use in preflight pilot briefings.

Pilots should review this information and acquaint themselves with routes located along planned flight paths and in the vicinity of airports from which they operate. Flight instructors, and flight schools especially, should be familiar with the routes that traverse their normal areas of operation. (Obtain the chart and post the information in the flight planning area.) Figure 4-5 contains an excerpt from an area planning, military training routes chart. Features include:

- Major airports
- NAVAIDs
- Flight service stations
- Military training routes
- Route altitudes
- Route hours of operation
- Special use airspace

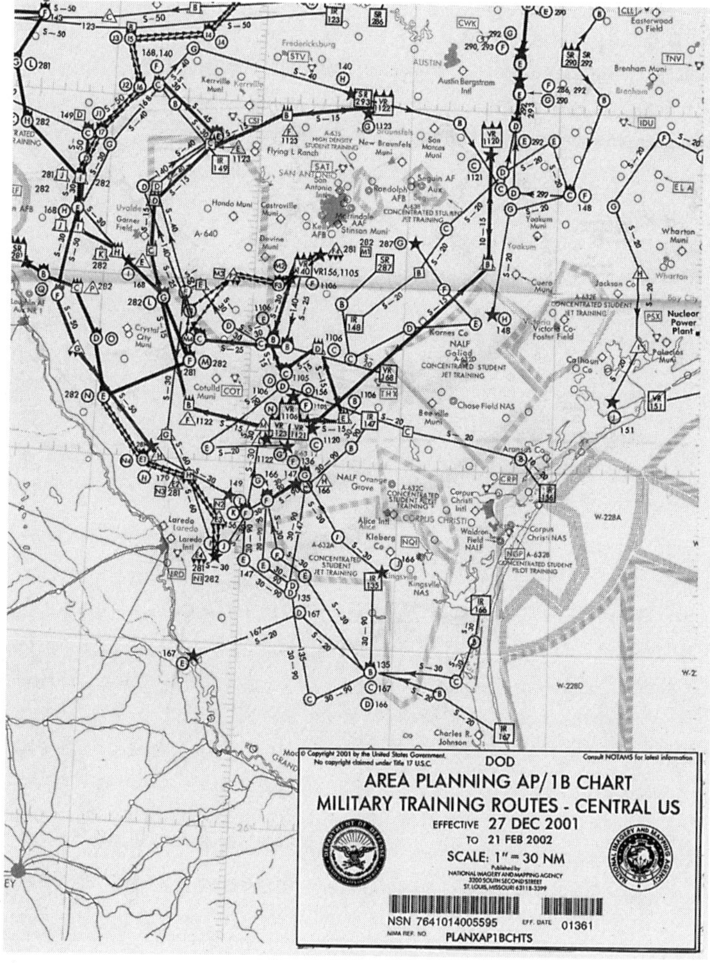

4-5 *Military training route charts provide a detailed visual depiction of low-level, high-speed routes.*

- Nuclear power plants
- Radioactive waste sites
- VFR helicopter refueling tracks
- Index to tactical pilotage charts

In addition to depicting IR and VR routes, the chart includes slow-speed, low-altitude training routes (SRs). These routes are used for military air operations at or below 1500 ft at speeds of 250 knots or less. Information about MTR activity is available from an FSS. Upon request, controllers at an automated FSS will provide information on military training routes along the pilot's route. Please provide the controller with the specific route numbers. Nonautomated FSS will normally have MTR route information only within 100 miles of the FSS's area. Because the area of MTR activity provided by a nonautomated FSS is limited, pilots should routinely request this information en route. Additional information about MTRs is published in a DOD FLIP for North and South America, which is covered in Chap. 9.

Until recently there have been no accidents involving military and civilian aircraft on these routes. The record is extremely good. This is in great part due to the vigilance of civil and military pilots and air traffic control. Like so many things in aviation, knowledge is a significant risk reducer. Information on these routes—location, times, and altitudes—is readily available, if we choose to use it.

Miscellaneous NIMA charts

Aerospace planning charts (ASCs) consist of six charts, at each scale, with various projections that cover the world. Chart scales are 1:9,000,000 and 1:18,000,000, with a sheet size of approximately 58 × 42 or 21 × 29 in, respectively. These charts are designed for wall mounting, and are useful for general planning, briefings, and studies. Charts do not contain contours or gradient tints, or aeronautical information. Cities of strategic or economic importance, major towns, transportation networks, international boundaries, prominent landmarks, and major lakes and rivers are shown.

Standard index charts (SICs) are graphics with index overprints for the major aeronautical chart series; sheet size is 28

× 48 in, covering the world, with a scale of 1:35,000,000. SICs are available for the following chart series:

- Global navigational charts
- Jet navigation charts
- Operational navigational charts
- Tactical pilotage charts

Canadian VFR charts

Canada produces and distributes its own set of pilotage charts. Similar to those used in the United States, these charts consist of world aeronautical charts, VFR navigation charts, and VFR terminal area charts. In addition, aeronautical planning, North Atlantic plotting, polar plotting, Canada–Northwestern Europe plotting, and Canada plotting charts are available.

Canadian WACs use the Lambert conformal conic projection and have a scale of 1:1,000,000. These charts serve the requirement of visual navigation for medium-speed, medium-range operations.

Canadian VFR navigation charts (VNCs) are equivalent to United States sectionals. VNCs use the Lambert conformal conic projection, with a scale of 1:500,000. They serve the requirements of visual navigation for low-speed, short- and medium-range operations, and are suitable for basic pilotage and navigational training. A special VNC has been developed for the Alaska Highway, and covers this route from Fort Nelson, Canada, to Northway, Alaska.

Canada also produces four VFR terminal area charts (VTAs)—for Montreal, Toronto, Winnipeg, and Vancouver—using the transverse Mercator projection, with a scale of 1:250,000, equivalent to United States TACs.

For details on aerial coverage and other Canadian chart information, visit the Canadian Map Office Web site. The Internet address is contained in Appendix B.

Topography and obstructions

Canadian charts display relief as contour lines, shaded relief, and color tints. Green color indicates flat or relatively level terrain, regardless of altitude above sea level. Significant elevations are depicted as spot elevations, critical elevations, and maximum elevation figures; MEF on Canadian charts indicates the highest terrain elevation plus 328 feet, or the highest known obstruction elevation, whichever is higher.

Hydrography and culture symbols are similar to those of United States charts. Obstructions are also indicated in the same manner, with obstructions 1000 ft AGL or higher shown with a larger symbol. Obstruction elevation in feet above sea level (ASL) appears above the height in feet AGL, which is enclosed with parentheses. Known obstructions 300 ft or higher and known significant obstructions below 300 ft are shown. When two or more are in the same area, only the highest obstruction is shown. Obstructions are lighted unless labeled "Unlighted."

Navigational aids

Approved land airports in Canada having runways 1500 ft or longer are charted. Airport symbols might be offset for clarity of presentation. Airports with hard-surface runways are depicted by the runway layout. When the use of a particular radio frequency is mandatory, the airport name is followed by the letter M. The appropriate frequency in the airport data is preceded by the letter A (airport traffic frequency). Other airport data and the availability of services is indicated in the same manner as United States charts. Airports where customs service is available are indicated by a broken-line box around the airport name.

Radio aids to navigation have the same general appearance as those used on United States charts. Heavy line boxes indicate services similar to an FSS with standard frequencies of 126.7 and 121.5 MHz. Other frequencies are shown above the box. At remote facilities the name of the controlling FSS appears in brackets below the NAVAID box. Control tower frequencies are not shown for all airports on Canadian WACs and VNCs, nor are they available in tabulated form as on United States charts. These frequencies must be obtained from other sources, such as VTAs or the Canada Flight Supplement, which is discussed in Chap. 9. Tower frequencies, along with detailed flight procedures, are contained in Canadian VTAs.

Navigational information

Canada uses standard International Civil Aviation Organization (ICAO) airspace class category system to describe airspace: A, B, C, D, E, F, and G.

In Canada, the base of Class A airspace varies from 18,000 ft MSL in the Southern Control Area, to FL230 in the Northern Control Area, and to FL280 in the Arctic Control Area. All controlled high-level airspace terminates at FL600.

Class B airspace is controlled airspace within which only IFR and controlled VFR (CVFR) flights are permitted. It includes all controlled low-level airspace above 12,500 ft above sea level or the minimum en route IFR altitude, whichever is higher. ATC procedures pertinent to IFR flights are applied to CVFR aircraft. Class B airspace terminates at the base of Class A airspace.

Class C airspace is controlled airspace within which IFR and VFR flight are permitted, but VFR flight requires a clearance from ATC to enter.

Class D airspace is controlled airspace within which both IFR and VFR flight are permitted, but VFR flights do not require a clearance from ATC to enter.

Class E airspace is airspace within which IFR are conducted, but VFR flights are not subject to control, as in United States Class E airspace.

In Canada, Class F airspace is of defined dimensions within which activities must be confined because of their nature, or within which limitations are imposed upon aircraft operations that are not a part of those activities, or both: United States special-use airspace equivalent.

Canadian special-use airspace is designated alert (CYA), danger (CYD), and restricted (CYR). Alert area activity is divided into one of the following:

- A acrobatic
- F aircraft test
- H hang gliding
- M military operations

- P parachute dropping
- S soaring
- T training

Altitudes are inclusive unless otherwise indicated. For example, CYA 125(A) to 5000, indicates an acrobatic flight alert area, active from the surface to 5000 ft MSL.

Navigational information consists of isogonic lines and values, local magnetic disturbance notes, aeronautical lights, airway intersection depictions, and VFR checkpoints. Most symbols are similar to their United States counterparts. Canadian charts depict VFR routes through mountainous areas. The route is indicated by a series of blue diamonds.

Case Study

In the late 1970s United States charts used this symbology to indicate recommended VFR routes through mountainous areas. Unfortunately, as in one case, a route that was not recommended (Kearsarge Pass, at 11,823 ft with nearby mountains well above 13,000 ft, through the rugged Sierra Nevada Mountains) found its way on the sectional chart as a series of blue diamonds. This led to great confusion and a hectic series of NOTAMs saying "No, we didn't mean that," and eventually all of these routes were removed.

Finally, Canadian charts cost more than double United States charts.

Other charts

Special-purpose and supplementary charts are also available. These consist of Gulf Coast aeronautical charts, the Grand Canyon visual flight rules chart, oceanic planning charts, United States VFR and IFR training charts, airport obstruction charts, state aeronautical charts, and Air Chart Systems charts.

Gulf Coast aeronautical charts

The United States Gulf Coast VFR aeronautical chart is designed primarily for helicopter operations in the Gulf of

Mexico, usually serving the offshore oil and gas interests. The chart shows the same onshore features as the WAC, covering the Gulf Coast and extending south to 26° 30′ north latitude. The chart is 27 × 55 in, which can be folded to the standard 5 × 10 in, has a scale of 1:1,000,000, and is revised annually. Features include, in addition to those shown on WACs:

- Offshore mineral leasing areas and blocks
- Oil drilling and production platforms
- High-density helicopter activity areas
- IFR GPS waypoint grid system

The waypoint grid system aids in the direct routing to oil platforms in the Gulf.

The IFR Gulf of Mexico vertical flight reference chart is designed for helicopter operations. This five-color chart depicts an IFR GPS waypoint grid system that aids in the direct routing to oil platforms in the Gulf. NAVAIDs and airport data are shown on this chart, along with special-use airspace and oil lease block information. Chart scale is 1:500,000 (1 in equals 70 nm). The chart is 45 × 36 in flat or 5 × 9 in folded; it is revised annually and printed front—west, and back—east.

Grand Canyon visual flight rules chart

NACO, in coordination with the FAA, has developed a Grand Canyon National Park chart. First published on April 4, 1991, this chart is designed to promote aviation safety and assist VFR navigation in this popular flight area. The Grand Canyon VFR aeronautical chart has a scale of 1:250,000, same as TACs, and will be revised as needed, probably once a year. The chart covers the procedures and restrictions required by Special Federal Aviation Regulation (SFAR) 50-2. One side of the chart is for noncommercial operations and the other side is for commercial air tour operations. Features include:

- SFAR operations below 14,500 ft MSL
- Flight-free zones, where aircraft operations are prohibited
- Corridors between flight-free zones
- VFR checkpoints

- Communications frequencies
- Minimum altitudes
- Navigational data

Oceanic planning charts

Oceanic planning charts (OPCs) are designed for transoceanic flights by pilots who do not have a navigator. The charts can be used in preflight and inflight planning, and rapid inflight orientation. Chart scales vary from 1:10,000,000 to 1:20,000,000, sheet size is approximately 17 × 11 in. Charts are available for the North Pacific and North and South Atlantic. International boundaries and continental outlines, along with selected radio aids, and no-wind equal-distance lines between select diversion airports are shown.

North Atlantic and North Pacific route charts are available from NACO. These multicolored charts are designed for the monitoring of oceanic flights by air traffic controllers. They may also be used by pilots for planning transoceanic flights. Charts are revised every 24 weeks.

The North Pacific route chart is a multicolored composite chart covering the North Pacific area. The chart has a scale of 1:12,000,000 (1 in equals 165 nm), sheet size is approximately 60 × 43 in unfolded. Four larger-scale area charts are in this series. The area charts cover four quadrants of the North Pacific for flight planning. Area charts have a scale of 1:7,000,000 (1 in equals 96 nm) and unfolded are the same size as the composite chart. Features include:

- Selected ATS routes
- NAVAIDs and reporting points with geographic coordinates
- International boundaries
- Russia and Russia-dominated areas
- U.S.-Russia Convention Line of 1867
- Buffer zones and nonfree flying areas
- Air defense identification zones
- International dateline
- Aerial refueling tracks

- Special-use airspace
- Airports of entry
- Mileage circles

The North Atlantic route chart is available in two sizes. Full size the scale is 1:8,250,000 (1 in equals 113.1 nm). The chart is 45 1/2 × 32 1/2 in. A half-size version is also available with a scale of 1:11,000,000 (1 in equals 151 nm), unfolded 29 × 21 in and folded standard 5 × 10 in. Features include:

- Selected ATS routes
- Oceanic control areas
- NAVAIDs and reporting points, with geographic coordinates
- North Atlantic/minimum navigation performance specifications area
- Air defense identification zones
- Airports of entry
- Flight information region (FIR) boundaries
- Shorelines
- International boundaries
- Special-use airspace

NIMA produces oceanic planning charts for the North Pacific Ocean (a two-chart series), and North and South Atlantic Oceans (a three-chart series). These charts are designed for aircraft making transoceanic flights whose aircrew excludes a navigator. Used for preflight and inflight planning, they allow rapid inflight orientation and assessment of position for determining alternate routes to diversion airfields should the situation warrant.

United States VFR and IFR training charts

NACO has two charts for training purposes only. One consist of the Seattle sectional aeronautical chart, the other the L-27/L-28 enroute low altitude chart. These are nothing more than copies that are specifically: "NOT TO BE USED FOR NAVIGATION." The first editions of these charts were issued in December 27, 2001. Training charts are designed to aid in teaching air navigation to student pilots and used to teach map-reading skills. They may be used for classroom training where a large number of charts are required. Since they are not updated on a regular basis, their cost is significantly less than their operational counterparts. These charts will be updated when significant specification changes take place.

Airport obstruction charts

NACO publishes airport obstruction charts (OCs), with a scale of 1:12,000, that graphically depicts 14 CFR Part 77, "Objects Affecting Navigable Airspace." OCs provide data for computing maximum takeoff and landing weights of civil aircraft, for establishing instrument approach and landing procedures, and for engineering studies relative to obstruction clearing and improvements in airport facilities. Features include:

- Airport obstruction information
- 14 CFR Part 77 surfaces
- Runway plans and profiles
- Taxiways and ramp areas
- Air navigation facilities
- Selected planimetry

(Planimetry shows man-made and natural features, such as woods and water, but does not include relief.)

State aeronautical charts

Many states publish aeronautical charts that cover the area within their boundaries, usually based upon the WAC scale of 1:1,000,000. The reverse side of the chart often contains specific airport information, airport diagrams, and other useful aeronautical or tourist information. Charts might be available from local distributors of aeronautical charts or more often from a state's transportation department or affiliated aeronautics agency.

Air Chart Systems

Howie Keefe developed Air Chart Systems charts to minimize the excessive work and cost of current charts. The VFR group includes VFR Enroute, VFR Terminal Atlas, and Loran/GPS Navigator Atlas. A feature of the service is frequent—56 day—updates. The system allows the VFR pilot to have en route altitude, distances, and reporting points otherwise available only on IFR charts. Addresses and telephone numbers for this service are contained in Appendix B.

5

Enroute charts

New radio navigation aids were developed for enroute position finding during the Second World War. The Coast and Geodetic Survey started developing a series of radio direction finding charts in 1939 to cover the United States. The *long-range navigation* (LORAN) system was established on the East Coast in 1941. New charts were developed to accommodate this new navigation system and a new series of world aeronautical chart (WAC) scale maps for the western hemisphere were developed, and completed for the rest of the world in 1943. Additional series for world planning and world long-range charts were initiated. In 1942, the first of a new series of instrument approach and landing charts was distributed by the geodetic survey.

From the middle to late '40s, the United States Air Force published *Instrument Let Down* publications. These procedures were bound volumes consisting of four charts to each page produced by the Coast and Geodetic Survey. A new series of radio facility charts was introduced by the survey in 1947, superseded 2 years later by a series of 59 standard radio facility charts covering the entire United States.

The Coast and Geodetic Survey introduced a new family of aeronautical charts in the early 1950s to simplify high-speed jet and transport navigation. The series included planning, radio facility, approach and landing, and visual charts. Figure 5-1 illustrates an approach and landing chart for this period. It is a low-frequency radio range approach for Edwards Air Force Base in California. These included jet navigation charts (JNCs) for visual flying and the issuance of new experimental approach and landing charts for instrument operations. With the new format that permitted

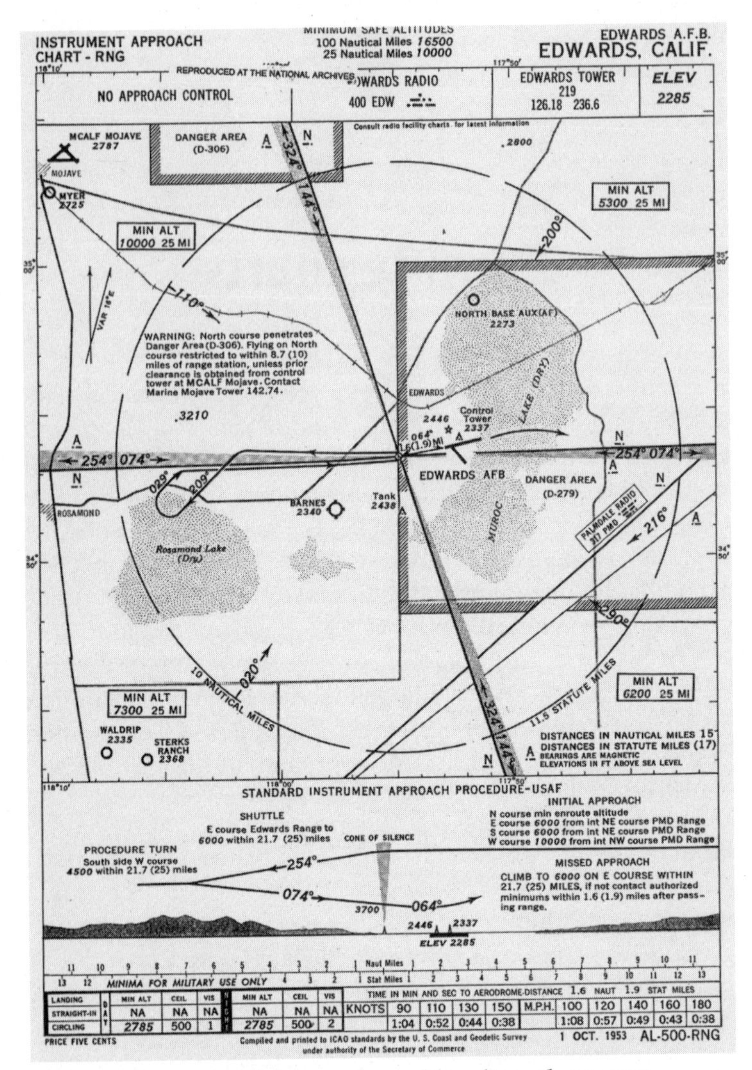

5-1 *A new series of approach and landing charts was introduced in the early 1950s to simplify navigation. Most were based on the low-frequency radio range, as in this approach for Edwards Air Force Base in California.*

two procedures to be printed on one side of the sheet, it was hoped that the more than 1100 instrument approach charts could be reduced to approximately 400. The smaller-size sheets were also easier to handle in the aircraft. Specialized charts were also developed and tested: operational navigation charts (ONCs), global navigation charts (GNCs), and various long-range navigation charts (loran, CONSOL, and CONSOLAN).

The radio facility charts proved unsatisfactory for jet aircraft flying at speeds faster than 500 knots. Consequently, in 1953, the Aeronautical Chart and Information Center introduced a new series of experimental radio facility charts that covered an area 750 × 250 nm and folded to 4½ × 9 in for convenient handling.

Area navigation prompted the National Oceanic and Atmospheric Administration (NOAA) to produce a series of high-altitude RNAV charts crisscrossing the United States with RNAV jet routes. These proved to be of limited value, however, and were subsequently discontinued.

Low- and high-altitude instrument enroute charts for the contiguous United States and Alaska are published by the National Aeronautical Charting Office (NACO). Enroute charts for other parts of the world are produced by the National Imagery and Mapping Agency (NIMA), and available through NACO. Canada also publishes enroute charts, as do several private vendors of instrument charts. Most of the information is for civilian aviation, but because NACO publications also serve the military, various NAVAIDs and services might not apply to civil operations.

Terminology and symbols

Terms and symbols discussed in this portion of the chapter apply to enroute planning charts, enroute low- and high-altitude charts, and area charts. Other IFR products (departure procedure, standard terminal arrival route, and approach and landing charts) use similar symbols in various colors. Since NACO charts are used by the armed services, some terms and symbology apply only to military operations. Finally, recall presentation limitations and problems of the cartographer as they apply to instrument charts.

Beginning in 1993 the color format of IFR charts changed. Prior to this time charts, were printed in two colors—blue and brown. The FAA and NOAA, in response to a request by the Aircraft Owners and Pilots Association (AOPA), has adopted a four-color scheme for IFR charts used in the United States. The new chart colors make the charts more readable, especially at night in low or red light; and, therefore, improve pilot comprehension and safety.

In general the colors depict the following information:

- *Blue*—airspace and airports
- *Black*—airway information and VHF NAVAIDs
- *Brown*—low frequency NAVAIDs, MTRs, and MOAs
- *Green*—coast line

Airports

Airport symbols on instrument charts are similar to those on visual charts, except that runway layout is omitted. On enroute low-altitude charts, all active airports with hard-surface runways—in Alaska hard or soft—of 3000 ft or more are depicted, along with all active airports with approved instrument approach procedures regardless of runway length or composition. On United States high-altitude charts, depicted airports have a minimum hard surface runway of 5000 ft—in Alaska hard or soft surface of 4000 ft.

Airports shown in blue or green have published instrument approach procedures, those in brown do not. The difference between airports shown in blue and those shown in green applies to military operations. Airports shown in blue have an approved Department of Defense (DOD) instrument approach procedure, those shown in green do not.

Airports are identified by their name; in the case of military airports, the abbreviated letters AFB (Air Force Base), NAS (Naval Air Station), NAF (Naval Air Field), MCAS (Marine Corps Air Station), or AAF (Army Air Field) also appear. Parentheses around the airport name means no military landing rights are available. That is, the military must obtain prior permission to use these airports. As in VFR charts, if the airport name is enclosed by a solid box, 14 CFR Part 93 special requirements apply. Airport information is similar to that used on visual charts. Runway length is the length of the longest active runway including displaced thresholds, excluding overruns. Runways that are not hard surface have a small letter s following the runway length, signifying a soft surface. An L following the elevation means that runway lights are on during hours of darkness. A circle around the L indicates that lighting is less than continuous. Airports within Class C or Class D airspace are indicated by the letter C or D within a small box following the airport name. Parentheses around the letter A (A) indicate that Automatic Terminal Information Service (ATIS) is available, along with the appropriate frequency. Airport symbols may be displaced to accurately depict the location of enroute navigational aids.

Radio aids to navigation

The depiction of radio aids to navigation (NAVAIDs) is similar to visual charts. Very high frequency (VHF) and ultrahigh-frequency (UHF) NAVAIDs [VORs, TACANs, and UHF nondirectional beacons (NDBs)] are shown in black. (UHF NAVAIDs are generally used only by the military.) Low/medium-frequency (LF/MF) NAVAIDs (compass locators and aeronautical or marine NDBs) are shown in brown. Notice that, in addition to the depiction of VORs, VOR/DMEs, and VORTACs, there is a TACAN-only symbol, illustrated in Fig. 5-2. NAVAID orientation to magnetic north may be illustrated by a magnetic tick, as shown in the TACAN example of Fig. 5-2. (With the military's use of GPS, TACAN undoubtedly will be the first ground-based navigation system to be decommissioned.) TACAN-only facilities are normally located on or in the vicinity of military bases.

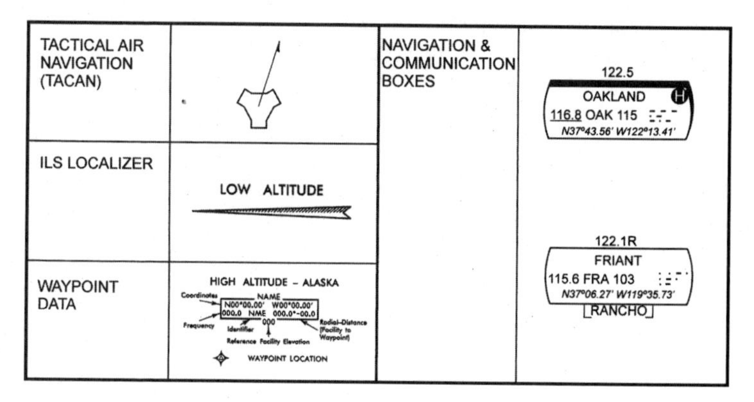

5-2 *Enroute chart airport and navigation aid symbols are similar to those used on visual charts.*

Without a TACAN receiver, civil pilots will normally not be able to utilize these facilities. However, civil pilots would be able to obtain DME information from a TACAN by tuning the DME receiver to the paired VOR frequency for the TACAN, contained in the *Terminal Procedures Publication* (TPP) and the *Airport/Facility Directory* (A/FD). In extremely congested areas, the NAVAID box will contain only the three-letter identifier; the complete NAVAID box will appear in a less congested area of the chart.

Instrument landing system (ILS) localizer courses that have an ATC function—used with another NAVAID to establish an airway intersection—are depicted as shown in Fig. 5-2. The feathered side indicates the blue sector.

VORs, DMEs, and TACANs are classified by their standard service volumes (SSVs). This determines the distances and altitude a particular NAVAID can be relied upon for accurate navigational guidance. Nondirectional radio beacons (NDBs) are classified according to their intended use. SSVs for VHF/UHF NAVAIDs are shown in Fig. 5-3. This information is of particular importance to pilots using VORTAC area nav-

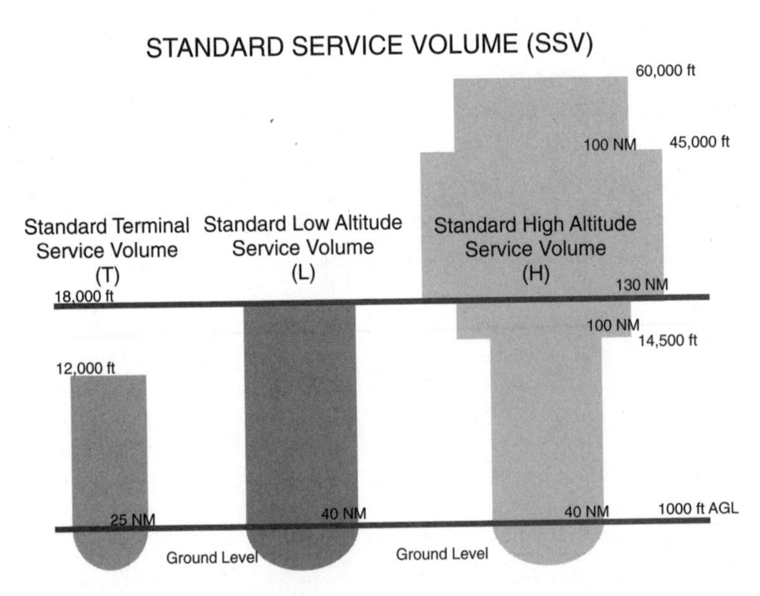

5-3 *Standard service volumes prevent signals from one NAVAID interfering with aircraft navigation from other NAVAIDs.*

igation (RNAV). For reliable navigational signals RNAV waypoints must be within the parameters of the SSVs shown in Fig. 5-3, as well as any restrictions to the NAVAID published in the A/FD and NOTAMs.

Over the last 30 years, the FAA has periodically proposed the decommissioning of all nondirectional radio beacons. With the evolution of GPS, this 1930s technology will certainly follow the low-frequency radio range into obscurity. But, the FAA will not put off to tomorrow what it can put off to the day after tomorrow. So, at least for a while, this navigational aid will be around. Table 5-1 lists SSVs for NDBs according to their classification.

NAVAID and FSS communication boxes on enroute charts are similar to those on visual charts, but in a slightly different format. This is illustrated in Fig. 5-2. A shadow box indicates an associated FSS with the same name as the NAVAID. Standard FSS frequencies are available. (Canadian standard FSS frequencies are 126.7 and 121.5 MHz.) In Fig. 5-2 the Oakland FSS has standard frequencies—122.2 and 121.5. In addition 122.5 is available at the facility. When an FSS is not associated with a NAVAID it is depicted by a shadow box with name of the facility, as in visual charts. When an FSS RCO is colocated with a NAVAID, the controlling FSS name appears below the NAVAID box. In Fig. 5-2 "Rancho" radio has a receiver-only at the Friant VORTAC on 122.1 and can transmit over the VOR frequency of 115.6. For aircraft equipped with coordinate navigation systems, the latitude/longitude of the facility is incorporated in the NAVAID box.

Table 5-1 NDB standard service volumes

Class Designator	Distance, nm
Compass locator	15
MH	25
H	50*
HH	75

*Service ranges of individual facilities may be less than 30 nm. Restrictions to service volumes are contained in the *Airport/Facility Directory*.

Terminal NAVAIDs are indicated by (T) following the NAVAID name. [On high-altitude charts, low-altitude NAVAIDs are designated by (L). The omission of a classification means the NAVAID is class H.] The letter (Y) shows that the TACAN receiver must by placed in the Y mode to obtain distance information—military use only. Overprinted data in the identification box indicates the facility might not be operating normally. That is, the facility might be shut down or has not yet been commissioned. The underlining of a channel means voice communications are not available.

Waypoint data is provided in two formats. First the latitude/longitude coordinates are provided as illustrated in the Fig. 5-2. VORTAC RNAV waypoints are also identified by the parent facility frequency, identifier, radial, and distance. Elevation of the parent reference facility is below the waypoint data box.

Airspace information

In this section we will discuss airspace information unique to IFR enroute low/high-altitude contiguous United States and Alaska charts. (Symbology already mentioned will not be repeated.) Information strictly related to military operations is also, for the most part, omitted—along with symbology that is evident. This will allow us to concentrate on civil, IFR operations. Figure 5-4 contains airspace information exclusive to IFR enroute charts.

VOR routes on low-altitude charts are designated *victor airways,* labeled with a V (V31 pronounced victor thirty-one), and depicted in black. VOR airways on high-altitude charts are called *jet routes* and labeled with a J (J5 pronounced jay five, not jet five). Jet routes are based upon VOR or VORTAC NAVAIDs and are depicted in black. In Alaska, selected segments of jet routes are based on L/MF NAVAIDs and are shown in brown.

Airways allow the pilot to file a route omitting intermediate fixes. For example, a pilot filing from San Francisco to Santa Barbara can file SFO V25 RZS SBA (San Francisco victor twenty-five San Marcos direct Santa Barbara), omitting all

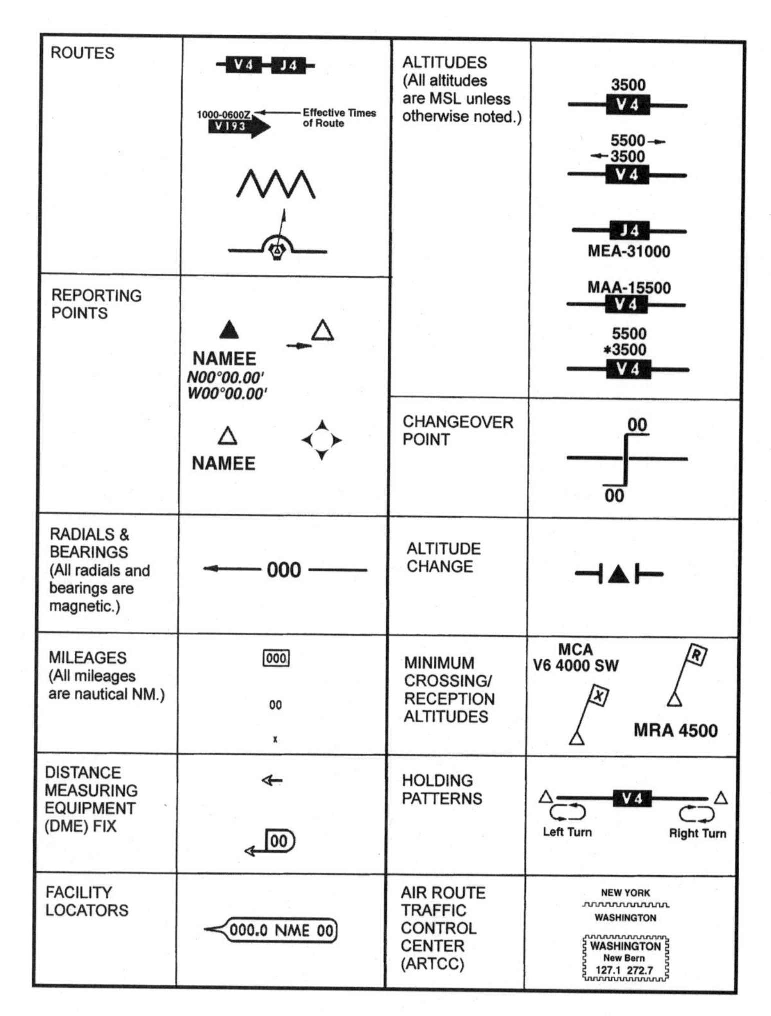

5-4 *On enroute charts, radials and bearings are magnetic, distances nautical, and altitudes feet above mean sea level, unless otherwise noted.*

intermediate intersections and NAVAIDs. L/MF airways are labeled B7 (blue seven), A15 (amber fifteen), G30 (green thirty), or R10 (red ten). These were the original radio navigation routes and not surprisingly were known as *colored airways*. Used mainly for oceanic routes, air traffic service routes are normally labeled with a letter and a number, similar to L/MF airways (R464, B592, and the like). VHF/UHF oceanic airways are depicted in black, LF/MF airways in brown. Single-direction routes are shown with an arrow, along with effective times. When routes are temporarily unusable, they are covered with a black zigzag line, as illustrated in Fig. 5-4. Occasionally, jet routes overlie a NAVAID that is not part of the route. In such cases the jet route centerline bypasses the facility that is not part of that specific route. This is illustrated in Fig. 5-4. This could be important in flight planning. Should a pilot file the route and include the bypassed facility, the ARTCC computer will reject the flight plan.

Reporting points consist of open or solid triangles or waypoint symbols. These are illustrated in Fig. 5-4. The reporting point will have the name of the associated NAVAID or a five-letter identifier (NAMEE). Solid triangles are compulsory and open triangles are noncompulsory reporting points. As required, coordinates are shown for compulsory, offshore, and holding fixes. Offset arrows indicate which facility forms the reporting point. On high-altitude charts the contraction NR indicates mandatory reporting is not required at the next compulsory reporting point.

VOR radials are magnetic, depicted outbound or from the facility; LF/MF bearings are magnetic, depicted inbound or to the facility. All distances are nautical miles and altitudes in feet above mean sea level (MSL), unless otherwise noted. Airway data, such as identifications, bearings or radials, mileage, and altitudes on low and high enroute charts are shown aligned with the airway and in the same color as the airway.

Refer to mileages in Fig. 5-4. The letter x shows a mileage break. These are often used for dogleg airways where a named intersection is not present. Pilots must use caution to include both distances along the airway for total mileage calculations.

Distance measuring equipment (DME) fixes are denoted by an arrow and sometimes a distance. The arrow denotes the DME distance is the same as the airway or route mileage. When a mileage is included with the symbol, it denotes that the DME distance is the mileage within the DME symbol. Both are illustrated in Fig. 5-4.

The minimum enroute altitude (MEA) indicates the minimum published altitude that assures acceptable navigational signal coverage, meets minimum obstruction clearance requirements between fixes, and ensures required radio communications. In Fig. 5-4 the MEA on V4 is 3500 ft. MEAs are sometimes different for opposite directions along an airway because of rising or lowering terrain. This is illustrated in Fig. 5-4. The MEA to the east is 5500 ft, to the west 3500 ft. When the MEA for jet airways is above 18,000 ft, it is specified, as illustrated in Fig. 5-4.

Additionally, where required the maximum authorized altitude (MAA) is published. Maximum authorized altitude (MAA) is the highest altitude, for which an MEA is designated, where adequate NAVAID signal coverage is assured. For example, an MAA will be established for route segments where interference from VOR signals on the same frequency prevent reliable navigation. In the example, it is 15,500 ft.

Minimum obstruction clearance altitude (MOCA) meets obstruction clearance criteria between fixes, but only assures navigational signal coverage within 22 nm of the NAVAID. Where established, the MOCA is shown directly below the MEA and is identified by an asterisk. The designation of a MOCA indicates that a higher MEA has been established for that particular airway or segment because of signal reception requirements. When no MOCA appears, the MEA and MOCA are considered to be the same.

Some airways have an MEA gap indicating that NAVAID signal coverage will be lost for a portion of the flight. Specific criteria exist for an airway or route segment to be designated with a navigational gap. The gap cannot exceed a distance that varies directly with altitude from zero at sea level to a maximum of 65 nm at 45,000 ft. Not more than one gap can exist, and the gap will not normally occur at a turning point. When a gap occurs, it will be identified by distances from the navigation facilities.

Changeover points (COPs) tell the pilot where to change from one navigational facility to the next. These COPs assure continuous reception of reliable navigation signals at the prescribed MEA. Where frequency interference or other navigation signal problems exist, the COP will be at the optimum location, taking into consideration signal strength, alignment error, or any other known condition that affects reception. Where signal coverage overlaps, the COP will normally be designated at the midpoint. The COP symbol will be omitted when the COP is at the midpoint of an airway segment.

When a change in MEA occurs at other than a NAVAID, an altitude change symbol is used. This is shown in Fig. 5-4 and appears as a T at the airway end.

Minimum crossing altitude (MCA) points out an associated NAVAID or intersection that cannot be crossed below a specific altitude. MCAs are established where obstacles prevent a pilot from maintaining obstacle clearance during a normal climb to a higher MEA after the aircraft has passed the point where a higher MEA applies. Standards for determining MCA are based on the following climb rates, and computed from the flight altitude:

- Sea level through 5000 ft: 150 ft per nautical mile.
- 5,000 through 10,000 ft: 120 ft per nautical mile.
- Above 10,000 ft: 100 ft per nautical mile.
- Some fixes have more than one MCA depending upon the direction of flight.

Minimum reception altitude (MRA) denotes the lowest altitude required to receive adequate signals to determine specific fixes. Often DME can be used to identify the fix, in which case the MRA would not apply because the off-airway facility would not be needed to establish the intersection. Pilots must use caution; if the aircraft is not DME equipped, the pilot might not be able to establish a required fix.

To simplify ATC instructions, where needed, holding patterns are depicted on enroute charts. In such situations, the controller will simply instruct the pilot to: "hold as published."

Air route traffic control center (ARTCC) boundaries and remote sector discrete frequencies are depicted on enroute charts. Within a communications box, the name of the center appears at the top, with the center's sector name in the middle. Specific frequencies are displayed at the bottom of the box (VHF for civilian use; UHF for the military).

VHF airways have a width of 4 nm each side of centerline to 51 nm. If the distance from the facility to the changeover point is more than 51 nm, the outer boundary of the primary area extends beyond 4 nm, diverging at an angle of $4^1/_2°$, as illustrated in Fig. 5-5. (This helps explain the shape of the Class E airspace between NAVAIDs on the visual chart in Fig. 2-1.) This area provides obstruction protection based on system accuracy. That is, where the NAVAID is within tolerance, the pilot is flying the centerline, and the navigation receiver is within required limits.

It is the pilot's responsibility to select altitudes that comply with obstruction clearance requirements for instrument flight along routes not in controlled airspace, or routes for which no specified minimum IFR altitude has been established.

The primary enroute obstacle clearance area extends from one navigational facility to the next. The minimum obstacle clearance over areas not designated as mountainous under 14 CFR Part 95 of the regulation, "IFR Altitudes," is 1000 ft above the highest obstacle. Because of the Bernoulli effect and atmospheric eddies, vortices, waves, and other phenomena associated with strong winds over mountains, steep horizontal pressure gradients develop in these regions. Downdrafts and

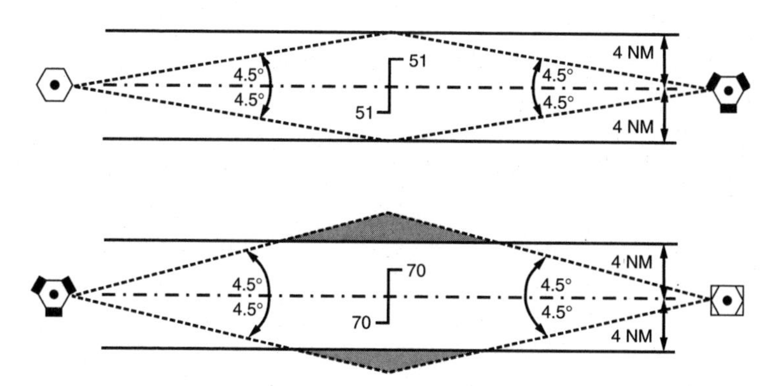

5-5 *Airway primary obstacle clearance areas extend 4 nm each side of centerline to 51 nm, then the area increases at an angle of 4.5°.*

turbulence are also prevalent under these conditions, which create significant hazards to air navigation; therefore, minimum obstacle clearance over terrain designated as mountainous in 14 CFR Part 95 is 2000 ft.

Case Study

Pilots planning low-altitude flight off airways must comply with minimum altitudes as specified by the regulations. The airway from Bakersfield to Santa Barbara in California has an MEA of 9000 ft. But, by checking terrain on the sectional, I was able to fly direct to the Fellows VOR, then the Gaviota VOR to Santa Barbara. This route was only slightly longer, but I was able to fly at 7000 ft. This takes ATC approval based on traffic.

Off-route obstruction clearance altitudes (OROCA) were added to all U.S. low-altitude enroute charts beginning in May 1995. OROCAs assist pilots using area navigation systems. OROCA is an off-route altitude that provides obstruction clearance with a 1000-ft buffer for nonmountainous terrain and a 2000-ft buffer for designated mountainous areas. OROCA provide an obstruction clearance for every quadrangle bounded by ticked lines of latitude and longitude, similar to MEFs on visual charts. Quadrangle size will vary, depending upon the scale of the chart, but the standard size is 1° × 1°. This altitude is provided for obstruction clearance only. It does not provide for NAVAID signal or communications coverage, and will not be consistent with altitudes assigned by ATC. Symbols are printed in brown. The first digit or digits represent thousands of feet and the following slightly elevated digit will represent hundreds of feet.

Recall that airports with Class C or Class D airspace are indicated in the airport data block. The lateral extent of Class B and Class C airspace, along with the Class C airspace Mode C requirement, are depicted on low-altitude charts. The difference between Class B and Class C airspace is differentiated by the outer blue circle. Class B airspace is designated by a solid blue line encircling the airspace. Class C airspace uses a dashed blue outer circle. Additionally, Class C airspace usually covers a smaller area than Class B.

Special-use airspace (SUA) below 18,000 ft MSL is shown on low-altitude charts; high-altitude charts show SUA above 18,000 ft. Open areas (white) indicate controlled airspace, shaded areas (brown) indicate uncontrolled, Class G airspace on low- and high-altitude charts. If an airway penetrates a restricted or prohibited airway, the airway may be subject to restrictions. In such a case, the airway is overlaid with a series of black dots.

Altimeter settings to be used are designated by the international codes QNH and QNE. QNH is the altitude above mean sea level displayed on the altimeter when the altimeter setting window is set to the local altimeter setting. This is the setting used in the United States for flying below 18,000 ft. QNE is pressure altitude. This is the altitude shown on the altimeter with the altimeter set to 29.92 in (1013.2 millibars). This is the setting used in the United States for flying at or above 18,000 ft.

Enroute charts use various cultural, hydrographical, and navigational and procedural information symbols. Cultural symbols identify international, convention, or mandate lines and dateline boundaries. Isogonic lines and values, time zones, and a translation of Morse code signals are provided. Shorelines are indicated by a green water vignette; smaller-scale area charts indicate a shoreline with a broken line. Match marks show chart overlap.

IFR enroute charts

IFR enroute charts consist of planning charts, low- and high-altitude charts, and area charts. IFR planning charts provide the IFR pilot with essentially the same information that VFR planning charts provide the VFR pilot. Low-altitude charts can be compared with sectionals, high-altitude charts with WACs, and area charts with TACs. Each is designed to provide enough detail to allow the pilot to operate safely in that particular environment.

Recall from Chap. 2 the discussion of NAVAID names and that NAVAIDs with the same name as an airport, and not collocated, were changed to prevent confusion. With the computerization of the ATC system a similar fate befell intersection names. Prior to this period intersections were named, like NAVAIDs, for geographical locations—usually towns and cities—adjacent to their location. They were abbreviated by using an alphanumeric code—for example, TWIN LAKES (4TW), GLENDALE (4GN), EL MONTE (4EM), and so on. Pilots had no need to know the codes because they were only used internally at air traffic control facilities. Now comes the early computer with its limited storage space. All intersections were changed to five letters. Since pilot self-briefing and flight plan filing systems (DUAT) came on line, pilots have been required to use the appropriate identifiers. With a limited number of five-letter combinations, this has led to some almost unpronounceable names. But some in the FAA do have a sense of humor.

Case Study

In northern California, a part of the Santa Rosa departure is the SNUPY intersection—named in honor of Charles Schultz, author of the Peanuts comic strip. Then in southern California we have the Spanish cow—ELMOO. Golfing enthusiasts will recognize some of their sport's parlance in the following intersections in the vicinity of Salt Lake City:

CHHIP
SPIEK
BOAGY
HELPR
DRYVE

Possibly the ultimate in intersection names can be found on the Portsmouth/Pease International Tradeport (PSM), Portsmouth, New Hampshire, GPS RWY 16 approach. Fixes form the initial approach fix to the missed approach fix are in the following order:

ITAWT
ITAWA
PUDYE
TTATT
IDEED

Eat you heart out, Sylvester and Tweety.

Planning charts

The VFR/IFR planning chart discussed in Chap. 4 supports IFR flight planning. Additionally, NACO publishes North Atlantic and North Pacific route charts. Indexes, in green, show coverage of associated IFR enroute low-altitude charts; indexes, in red, show coverage of associated sectional charts and include MEFs for the sectional coverage area.

Symbols used on the VFR/IFR planning chart are standard, with the following exceptions. Airports shown have a minimum 3000-foot hard-surface runway. Those in green have approved instrument approach procedures. Airports in blue, in addition to a low-altitude instrument approach, have military landing rights. Airports in brown do not have instrument approaches, and those in red are private. VHF/UHF NAVAIDs are depicted by standard symbols. Features include:

- Victor airways
- NAVAIDs
- Mileage
- Airports—3000 foot paved or with instrument approach
- Low-altitude enroute chart outlines
- ARTCC boundaries
- Special-use airspace below 18,000 ft

Enroute low-altitude charts

Enroute low-altitude charts are designed to provide navigation information for IFR flights below 18,000 ft MSL. Contiguous United States scale varies from 1:583,307 (1 in equals 8 nm) to 1:1,458,267 (1 in equals 20 nm). The scale of the individual charts depends on the amount of information displayed. In Alaska, because of the large area and relatively small amount of information depicted, charts have a scale of 1:2,187,402 (1 in equals 30 nm). All charts can be folded to the standard 5 × 10 in. Revision cycle is 56 days. Low-altitude enroute charts are labeled L with a number (L-13). Coverage for the contiguous United States is contained in Fig. 5-6. Coverage for Alaska is shown in Fig. 5-7. Features include:

- Low-altitude airways
- Limits of controlled airspace

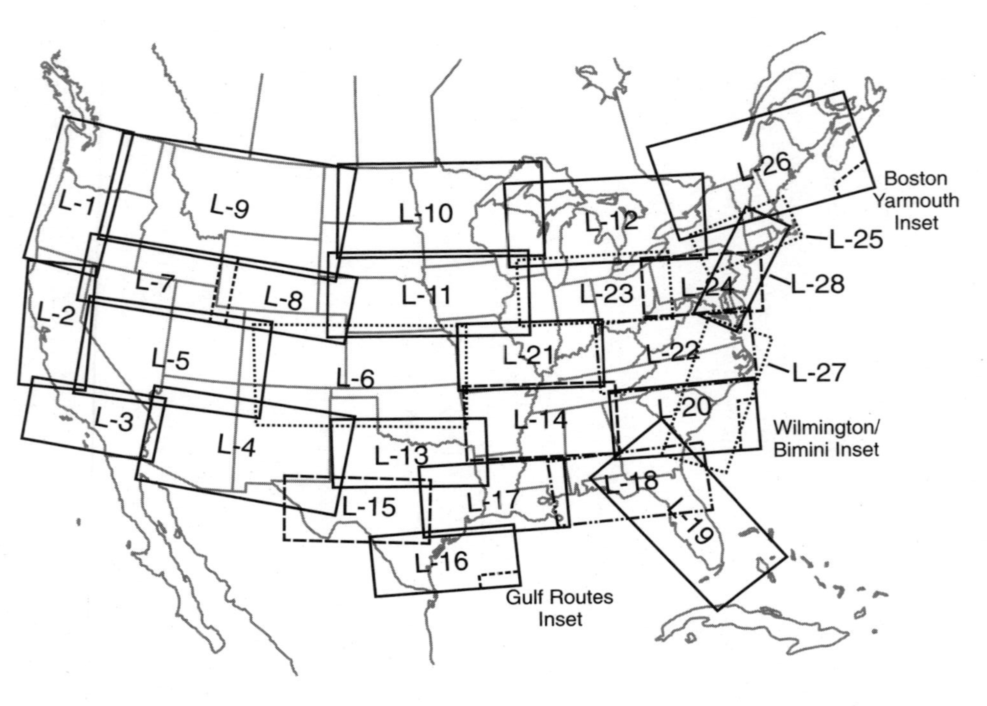

5-6 *The contiguous United States is covered by a series of 28 enroute low-altitude charts, printed on 14 sheets.*

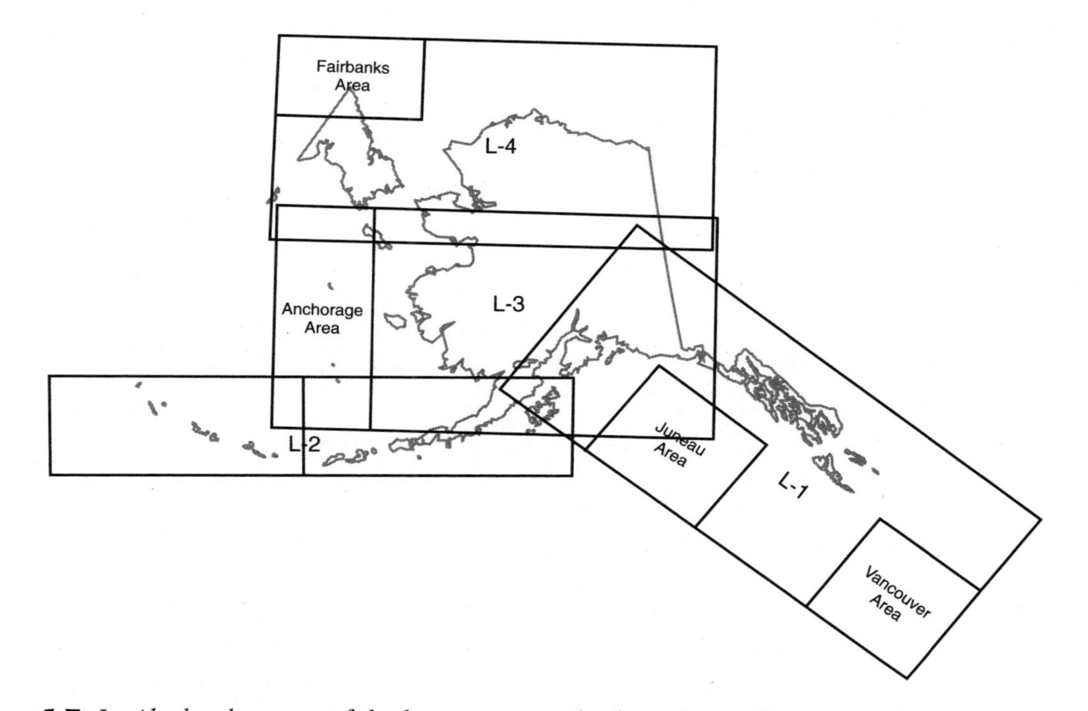

5-7 *In Alaska, because of the large area and relatively small amount of information, all four charts have the same scale.*

- Radio aids to navigation
- Selected airports
- Route altitude descriptions (MEAs, MOCAs, MCAs, OROCAs)
- Airway distances
- Reporting points
- Special-use airspace
- Military training routes
- Operational notes
- Adjoining chart numbers
- Outlines of area chart coverage

NACO will be adding terrain features to the area chart series.

Area charts are larger-scale representations of congested terminal areas. Scale varies from 1:364,567 (1 in equals 5 nm) to 1:583,307 (1 in equals 8 nm). NACO area charts are distributed on a single chart sheet. The chart folds to the standard 5 × 10 in, with a revision cycle of 56 days. End panels contain the standard air to ground communication frequencies listing for the areas covered. Area charts are available for the following locations:

- San Francisco
- Los Angeles
- Denver
- Minneapolis/St. Paul
- Dallas/Fort Worth
- Kansas City
- Chicago/Milwaukee
- Detroit
- St. Louis
- Atlanta
- Jacksonville
- Miami

Enroute high-altitude charts

Enroute high-altitude charts are designed to provide navigation information for IFR flights at and above 18,000 ft MSL.

Except to effect transition between the low- and high-altitude route structures, victor airways should not be flown above 18,000 ft. Pilots planning flights at or above 18,000 ft should file appropriate direct or jet routes for operations within this stratum. Charts for the contiguous United States have a scale of 1:2,187,402 (1 in equals 30 nm, except for the H-6 chart, which has a scale of 1 in equals 18 nm). Alaskan charts have a scale of 1:3,281,102 (1 in equals 45 nm). All charts fold to the standard 5 × 10 in, with a revision cycle of 56 days. High-altitude enroute charts are labeled H with a number (H-4). Figure 5-8 shows chart coverage for the contiguous United States. Be careful when selecting chart coverage. Chart H-1 has H-3 on the reverse side; chart H-2 has chart H-4 on the reverse side. Figure 5-9 shows coverage for the state of Alaska. Features include:

- Jet route structure
- Airspace information
- Special-use airspace
- Selected airports
- Radio aids to navigation
- Reporting points

The jet route structure is shown in black; terminal and low-altitude NAVAIDs are screened light gray. Airspace information is shown in blue, with SUA tabulated on the title panel. All airports with at least 5000 ft of hard-surface runway are depicted, except on Alaska H1 and H2 charts, which include hard-surface runways of 4000 ft. Airports displayed in blue and green have an approved instrument approach procedure, those in blue also have a high-altitude penetration procedure. Airports in brown have no approved instrument approach procedure. VHF NAVAIDs have frequency, identification, channel, and geographic coordinates displayed. Geographic coordinates are also provided for compulsory reporting points. MEAs are 18,000 ft, unless otherwise shown.

Using en route IFR charts

A pilot's first task is to ensure currency. NACO IFR charts are published every 56 days. Changes that occur between publication cycles are distributed in the form of a change notice

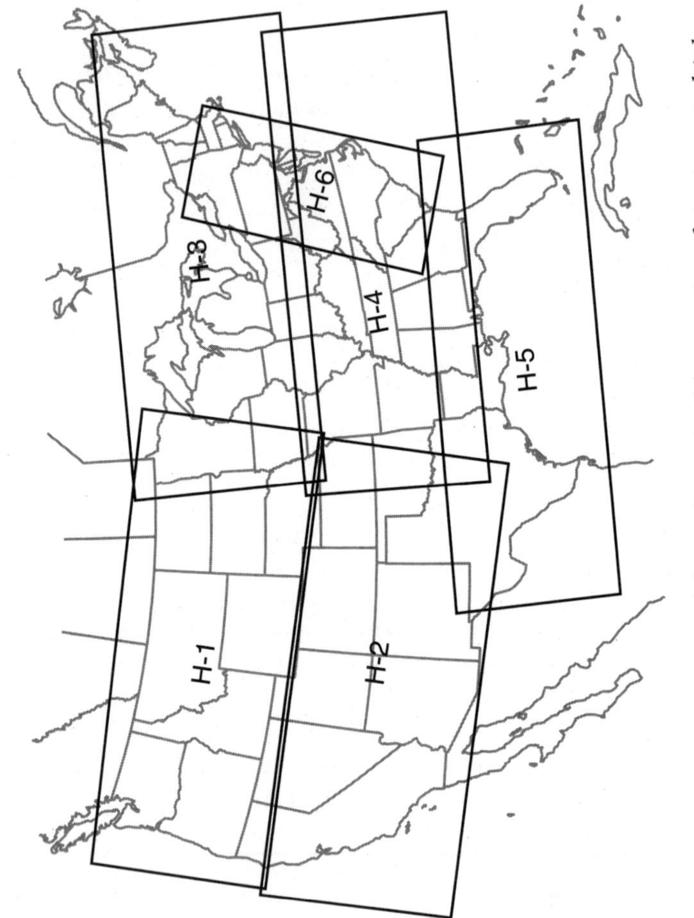

5-8 *The contiguous United States is covered by a series of six enroute high-altitude charts, printed on three sheets.*

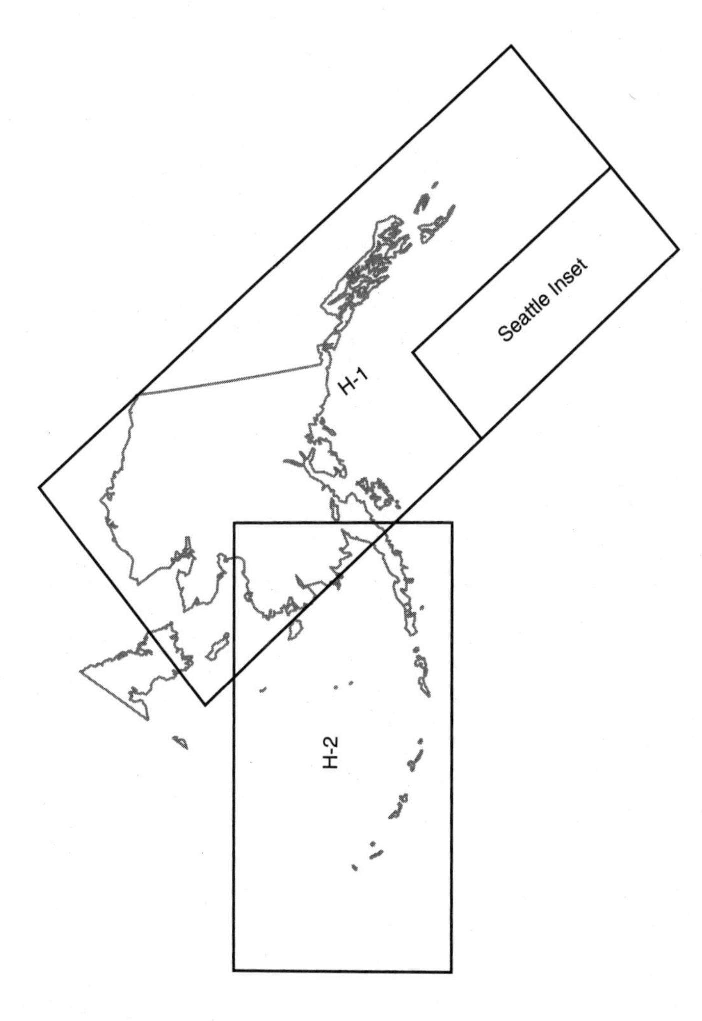

5-9 *Alaskan enroute high-altitude charts provide coverage for the state, west coast of Canada, and the northwest United States.*

volume, the *Notice to Airmen* (NOTAM) publication, and National Flight Data Center (NFDC) NOTAM. NAVAID and airport restrictions, changes, or outages might also appear in these publications, or be published as NOTAM (D)s. Publications, sources, and availability of products were discussed in Chap. 1. Pilots, especially those using NACO products, must be aware of these limitations.

Enroute low- and high-altitude chart cover panels identify the chart, provide effective times and date, and contain other useful information, such as applicable MTR tables and SUA data. Low-altitude charts provide a tabulation of airport and terminal communication frequencies. High-altitude chart cover panels contain discrete high-altitude flight watch frequencies for their area of coverage. Inside panels contain a brief chart legend. Chart margins provide a mileage scale, panel labels (panel E, F, G, and the like) with associated major city name, and pertinent remarks (the name of the next fix off the chart, or "Overlaps chart Nr L-5"). The A/FD references airport and NAVAID locations to their associated low- and high-altitude chart and panel. For example, the directory for the Santa Rosa, California, airport lists: H2G, L2G. This airport is located on the enroute high-altitude chart number 2, panel G (H2G), and the enroute low-altitude chart number 2, panel G (L2G).

Figure 5-10 contains a portion of the enroute low-altitude L-2 chart. Let's examine the airports, facilities, and airway structure between Santa Rosa and Mendocino in Fig. 5-10. Below is the depicted data:

Sonoma Co [D]*

125 (L) 51

(A) 120.55

The Sonoma County Airport is depicted in blue. The airport is served by a published low-altitude instrument approach procedure and an approved Department of Defense (DOD) approach procedure, or DOD radar approach, or both. The airport is located in part-time Class D airspace (D*). Airport elevation is 125 ft MSL, lighted for night operations—pilot controlled, and it has a runway of 5100 ft. ATIS is available on 120.55 MHz. The chart indicates that ATIS is continuous.

At airports with a part-time tower, such as Santa Rosa, the AWOS is broadcast over the ATIS frequency when the tower is closed.

To the northeast of Sonoma County are the Pope Valley and Angwin-Parrett Field airports, depicted in brown. Pope Valley is a private "Pvt" strip; both airports have pilot-controlled lighting (PCL) indicated by the circle around the L symbol. These airports do not have an approved instrument approach

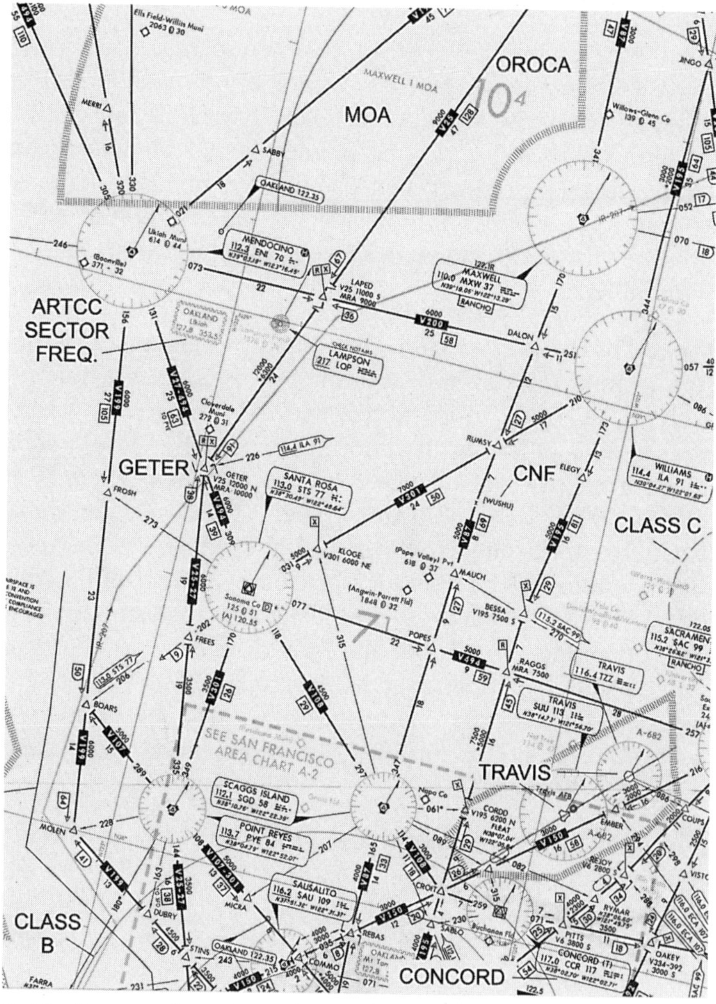

5-10 *Enroute low-altitude charts provide navigational information for flights below 18,000 ft.*

procedure, and the parentheses around the airport names indicate that military landing rights are not available.

Amid the NAVAIDs, notice that Santa Rosa is a VOR/DME and Mendocino a VORTAC, from the NAVAID symbols. Both NAVAID identification boxes contain standard information. The circle with the letter "H" in the upper left corner of the Mendocino NAVAID box indicates the availability of hazardous inflight weather advisory service (HIWAS). Remote FSS communications are not available over either VOR frequency, as indicated by the underlined frequency.

What is the difference between the open triangle at the center of the Santa Rosa NAVAID symbol and the solid triangle at the center of the Mendocino NAVAID symbol? Santa Rosa is a noncompulsory reporting point; whereas Mendocino is a compulsory reporting point. The passage of, and estimates to, compulsory reporting points is a required report to ATC. However, this requirement is waived for flying within a radar environment.

Notice the Lampson nondirectional radio beacon east of the Mendocino VOR. On top of the frequency box is a note to "CHECK NOTAMS." Also notice that the frequency is overprinted. This indicates an abnormal status. Therefore, pilots must check NOTAMs or the A/FD. Upon such a search a pilot would find no NOTAMs, either NOTAM (D) or published for this facility. However, a check of the A/FD reveals that the radio beacon is: "Out of svc indef." This emphasizes the necessity for pilots to check all available sources. (A pilot obtaining an FSS briefing would receive no NOTAM that the beacon is out of service. Even if the pilot requested the FSS controller to check the NOTAM publication, no NOTAM would be found. Why? The information is published in the A/FD.)

Travis AFB has two NAVAIDs named "TRAVIS." One is a VOR, frequency 116.4 MHz, identification TZZ. The other is a TACAN only (note the symbol), frequency Channel 113, identification SUU. Just south of Travis is the Concord VOR (CCR). After its name appears (T). This indicates that the CCR VOR is a terminal (T) NAVAID, only to be used within the terminal standard service volume (SSV).

The communication box above the Mendocino VOR indicates an FSS remote communications outlet. The controlling FSS is Oakland, with the single frequency of 122.35 MHz. While we're on communications, note the center sector frequencies to the southeast of the Mendocino VOR. This is Oakland Center's Ukiah sector, with frequencies of 127.8 and 353.5 MHz. IFR and VFR radar services are available from Oakland Center on these frequencies.

At times, there can be confusion about airway identification. Airways allow a pilot to file and fly, and ATC to assign, a single route element to describe courses to be flown that could cross the entire country. Often, more than one airway is designated by the same NAVAID radials. Note the airways that cross the GETER intersection northwest of Santa Rosa. The airway between Santa Rosa (STS) and Mendocino (ENI) is designated V494. After the GETER intersection, V27 also overlies the same radials as V494 (V27-494). Where did V27 come from? Airways V25 and V27 come from Point Reyes (PYE) to the GETER intersection. At GETER, V25 goes off to the north and V27 to Mendocino. Northwest of Mendocino, V27 and V494 separate. This is important during flight planning, especially for pilots using DUATs. The airway between STS and ENI is V494, not V27 or V27-494. Neither V27 nor V27-494 will be accepted during DUAT computer flight planning—it isn't accepted by the FSS computer either, but the FSS controller will coordinate with the pilot to correct the route.

Airway information interpretation is, for the most part, straightforward. The airway from STS to GETER is made up of the STS 309° radial, MEA 6000 ft, and distance 14 nm. GETER is the COP, as indicated by the COP symbol at the intersection. (The COP is normally located midway between navigation facilities, or at the intersection of radials or courses forming a dogleg. When the COP is not located at the midway point, aeronautical charts depict the COP location.) The airway from GETER to ENI is established using the ENI 131° radial, MEA the same, and distance 25 nm. (Someone is going say, "You mean 301° to ENI." A pilot would in fact select 301 on the VOR's omni bearing selector to fly *to* the ENI VOR. But, recall that all radials on the chart are *from* the facility. This convention is used to eliminate any confusion; well that's the idea anyway.)

The chart also provides total distances between NAVAIDs. The number 39 in a box just outside the STS VOR compass rose represents the total distance between STS and ENI. What about the 63 in a box just to the northwest of GETER? As the note states, 63 miles is the total distance between ENI and the PYE VOR, or the total distance between NAVAIDs on V27. Total distance between NAVAIDs on V25 is found along V25 north, away from GETER in the box showing 128 nm. Also note the MEA north of GETER on V25 is 12,000 ft. The number below, with the asterisk, *6300 ft, is the MOCA. Do not expect navigational signal coverage at the MOCA because this portion of V25 is well beyond 22 nm from the NAVAID that establishes the airway.

On V25 the airway contains an MEA change bar at GETER, in this case the change, represented by the "T," is quite significant. The MEA changes from 6000 to 12,000 ft.

Using VORs only, the GETER intersection is designated as any one of the four airway radials and the Williams (ILA) 226° radial. In order to establish an intersection, normally, the cross radial angular divergence must be at least 30°. Therefore, in order to hold at GETER, using VORs only, the ILA VOR must be used. However, to establish the COP a pilot could use time, based on the aircraft's ground speed.

GETER intersection has both MCA and MRA flags. GETER has an MCA on V25 of 12,000 ft northbound. The MRA flag warns that the lowest altitude to establish the intersection using the VOR radial that makes up V25 north of GETER is 10,000 ft. How else could a pilot establish this intersection? GETER can be determined using DME from STS (14 nm), ENI (25 nm), or PYE (38 nm)—the arrows on either side of GETER, or the DME arrow box south and north of the intersection.

Take a look a the small letter x on V87 between the RUMSY and POPES intersections. From our previous discussion we know the x indicates a mileage break. From RUMSY to the mileage break the distance is 7 nm, from the break to MAUCH intersection is 8 nm. Now the note (WUSHU) adjacent to the break. (WUSHU) is the computer navigation fix (CNF) for the mileage break and has no ATC function—the CNF allows pilots with data base navigation systems to more easily establish the mileage break.

Computer Navigation Fix

CNFs have been added to enroute, DP, STAR, and instrument approach procedures at mileage break points (turns) on airways, which formerly only contained the small letter x indicating a mileage break. CNFs are identified with a five-letter name enclosed in parentheses. Pilots using non-database navigation systems will continue to identify airway turns in the conventional manner. ATC will not request that aircraft hold at, or report these fixes. Pilots should not request routes via CNFs, nor refer to CNFs in communications or flight plans.

Enroute low-altitude charts depict the lateral boundaries of Class C and Class B airspace. These are both illustrated in Fig. 5-10. The western boundary of the Sacramento Class C airspace is shown, along with the northern edge of the San Francisco Class B airspace.

On the basis of the OROCA, what would be the minimum altitude for a flight from Santa Rosa to Willows-Glenn (Willows-Glenn is north of the Maxwell VOR in Figure 5-10.) OROCAs along the route are 7100 ft in the Santa Rosa grid and 10,400 ft in the Maxwell grid.

At the top of Fig. 5-10 is the Maxwell 1 MOA. IFR flight will not be cleared through this airspace while the area is in use by the military, unless ATC can provide positive separation between military and civilian aircraft. Toward the bottom of Fig. 5-10, surrounding Travis AFB is A-682. This Alert Area cautions pilots to extensive military operations in the vicinity of the military base. Also note that pilots in the vicinity of the San Francisco Bay are advised to consult the SAN FRANCISCO AREA CHART A-2 for operations in this area.

Figure 5-11 contains an example of an enroute high-altitude chart that covers the same general area as the enroute

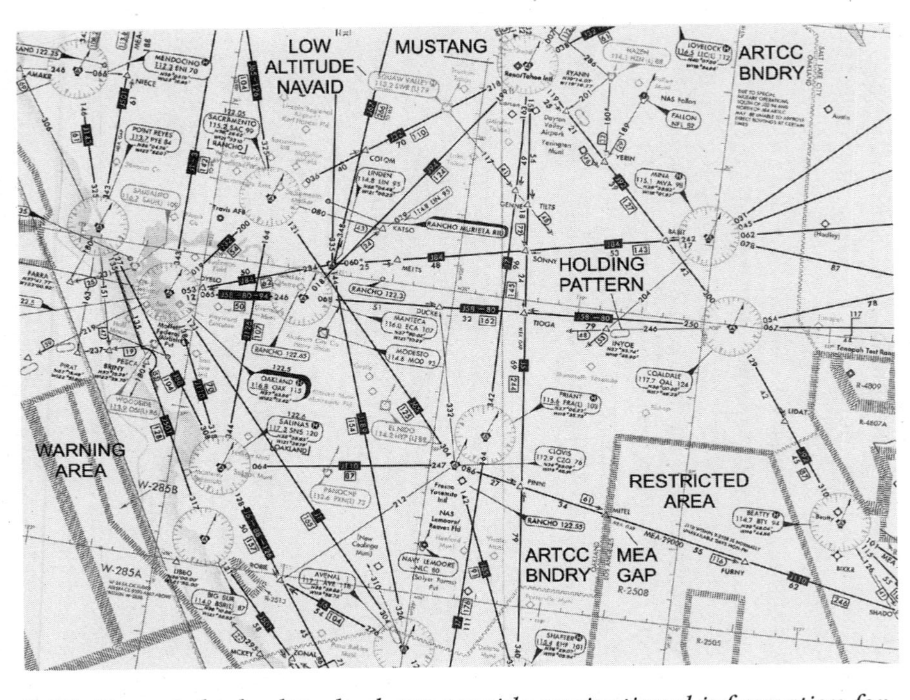

5-11 *Enroute high-altitude charts provide navigational information for flights at and above 18,000 ft.*

low-altitude chart in Fig. 5-10. Airport, NAVAID, and communication box symbols are common to both charts.

As with low-altitude airways, one radial may define more than one segment of a jet route. For example, east of the Oakland VORTAC the 065° radial is used to define J58, J80, and J94. J94 splits off at the Manteca (ECA) VORTAC and continues on to the Mustang (FMG) VORTAC. J58 and J80 continue eastward through the Coaldale (OAL) VORTAC, then diverge. It's important for pilots to study their routes and file the correct airways.

There are a number of low-altitude NAVAIDs depicted in Fig. 5-11—for example, the Squaw Valley VOR-DME. Notice the (L) following the station identification SWR in the NAVAID box.

East of the Clovis VORTAC is J110. This airway has a "MEA GAP" as well as a MEA of 29,000 ft. The airway penetrates restricted area R-2508. A note advises pilots that: "J110 within R-2508 is normally unavailable days MON-FRI." Coinciding with the western edge of R-2508 is a zigzag square line indicating the boundary between Oakland and Los Angeles ARTCCs. North of R-2508 is the INYOE intersection with a charted holding pattern.

Normally, pilots should file a departure procedure (DP) from the departure airport with a transition to a route or NAVAID within the low- or high-altitude airway structure to be flown. We would then plan direct, or victor, or jet routes to one of the transition fixes for the destination airport. Because the complexity of transitioning from the high altitude structure is often greater than that for the low-altitude structure, standard terminal arrival routes (STARs) are commonly used. DPs and STARs are specifically addressed in Chap. 6. Subsequent chapters also relate these low and high enroute charts to instrument approach procedure charts.

Although airway-to-airway routes can be filed (V28 V334), unless the computer can define the intersection, the flight plan will be rejected. It's always best to file the intersection (V28 ALTAM V334), which is required with some direct user access systems. This is certainly true for V25 and V27 along the west coast. Airway segments overlap in a number of places, thus requiring the pilot to specify a fix when transitioning from one airway to the other. On the other hand, one purpose of filing airways is to allow the pilot to file a route and omit intermediate fixes. Unfortunately, a few pilots still file V25 SNS V25 PRB V25 RZS. The purpose of the airway is to allow the pilot to file and ATC to issue a clearance "...V25 RZS...," omitting intermediate fixes.

Because of limitations of ATC computers, not all LOCIDs can be stored. A pilot using some direct user access systems can file ROM V485 HENER V186....DUAT will respond "Posting FIX REDDE for ZOA ARTCC adaptation purposes. Route of Flight...? ROM V485 REDDE V485 HENER...." This allows the Oakland Center computer to accept the flight plan. (Are ATC computers old? Well, some are still water cooled!)

Pilots flying aircraft equipped with approved area navigation systems might wish to file direct or point-to-point. The random route portion of the flight should begin and end over appropriate departure and arrival fixes. The use of departure procedure (DP) and standard terminal arrival routes (STAR) is recommended.

For VORTAC RNAV the route must be defined by using degree-distance fixes from appropriate navigational aids. As a minimum, one waypoint must be filed for each ARTCC through which the flight will be conducted. The waypoints must be located within 200 nm of the preceding ARTCC's boundary. This requirement is due to the limited storage capability of the ARTCC computers. If the computer does not recognize the fix, the flight plan will be rejected.

Case Study

A pilot wished to file RNAV direct from Hollister, California, to John Day, Oregon. When asked, he was unable to provide a waypoint. The FSS controller referred to a chart and acceptable waypoints were determined. The flight plan, as filed by the pilot, would have been rejected.

Pilots flying aircraft equipped with latitude/longitude coordinate navigation capability may file random routes at and above FL390, within the contiguous United States, using coordinates. Appropriate DPs and STARs should be used. After the departure fix, the pilot must include each turn point and the arrival fix for the destination. The arrival fix must be identified by coordinate and fix identifier. For example, OAK OAK5 OAK LIN 4044/11346 BVL BVL1 SLC. This route specifies the Oakland five departure to the Linden (LIN) VOR, direct to 4044/11346, which is the coordinate for the Bonneville (BVL) VOR. Bonneville is the entry fix for the BVL one arrival to the Salt Lake City (SLC) airport.

DUAT does not discriminate against pilots filing at any altitude for any direct route. Flight plans are accepted from any point to any point defined within the computer database. When necessary, DUAT inserts the latitude/longitude coordinates for the destination. Even when accepted by the computer, these flight plans don't work; someone, somewhere has to fix them, which results in delays to everyone.

Irate center supervisors have called demanding that the FSS include appropriate departure fixes. Well, that's not the FSS's job, it's the pilot's. On other occasions, especially when the control facility is extremely busy, pilots will be advised that their flight plan is incomplete and a clearance cannot be issued. Even when accepted by the computer, these flight plans don't work; someone, somewhere has to fix them, which causes in delays for everyone.

Communications failure

Charts, instrument and visual, become a pilot's lifesaver if radio communications are lost. The filed flight plan and route information published on either chart series becomes the pilot's air traffic controller. Figure 5-10, enroute low-altitude chart L-2, provides the basis for chart utilization during communications failure. Let's say we're just south of PYE and have been cleared direct PYE V-25 Red Bluff, we're at 6000 ft and have been instructed to expect 10,000 ft at FREES intersection. The aircraft is not DME or RNAV equipped. Radio communications cease; attempts to establish contact are unsuccessful. If in VFR conditions, or if VFR conditions are encountered, we will continue the flight VFR and land as soon as practicable.

Remember that operating under these conditions might unnecessarily, as well as adversely, affect other users of the airspace, because ATC might be required to reroute or delay other pilots in order to protect airspace for the aircraft without communications capability.

Case Study

A pilot on the ILS approach to Santa Rosa lost communications. Breaking out of the clouds the pilot proceeded to an uncontrolled airport, landed, and told no one. ATC had to sterilize the airspace, almost forcing a commuter jet to unnecessarily return to San Francisco because of low fuel. This pilot should have continued the approach and sorted things out on the ground at Santa Rosa. It is not the intent of "land as soon as practicable" to mean land as soon as possible. The pilot retains the prerogative of exercising judgment and is not required to land at another airport, an airport unsuitable for the aircraft, or to land only minutes short of the destination.

But, we're in IFR conditions. We are expected to continue by the assigned route in the last ATC clearance—if we're on a radar vector, by a direct route to a fix, route, or airway specified in the vector clearance, or by the expected route clearance issued by ATC, or in the absence of any of these, by the route filed in the flight plan. In our example the assigned route is direct PYE, then V-25.

We are required to fly at the highest of the following altitudes, for the route segment being flown:

- The altitude assigned in the last clearance.
- The minimum altitude for IFR operations (MEA).
- The altitude ATC has advised might be expected in a further clearance.

Based on these requirements we maintain 6000 ft until the FREES intersection, because that was the last assigned altitude and it is higher than the MEA (3500 ft) for this route segment. At FREES, we were instructed to expect 10,000, so we begin a climb to that altitude at that intersection. Now, what do we do at GETER? The MCA northbound on V-25 is 12,000 ft. Upon reaching GETER we are expected to enter a standard holding pattern and climb to cross GETER at 12,000 ft, and maintain that altitude, the MEA for that route segment. At the LAPED intersection, the MEA drops down to 9000 ft. Crossing LAPED, ATC expects us to descend to the last assigned altitude, or the MEA, whichever is higher, in this case 10,000 ft.

Let's say we're southbound on V-25 north of LAPED, assigned 9000 ft. From the MCA and MRA flags at LAPED we see the MRA is 9000 ft, no factor, but the MCA is 11,000 ft southbound. In this case we would be expected to enter a standard holding pattern at LAPED, climb and cross the intersection at 11,000, then continue the climb to the MEA of 12,000 ft.

Approach procedures during periods of radio communications failure are addressed in Chap. 7.

Supplemental IFR enroute charts

NIMA produces low- and high-altitude chart series that cover most of the world, consisting of enroute charts and supplements—FLIPs. Charts and FLIPs are available for sale from the NIMA on a one-time basis or annual subscription.

Low-altitude charts are printed back to back, and folded to the standard 5 × 10 in. They portray data required for IFR operations. Area charts are included for operations at selected terminals. The supplement is a bound book, approximately 5 × 8 in, containing an alphabetical IFR/VFR airport/heliport facility directory, airport sketches, and data required to support the enroute and area charts.

NIMA also produces a series of area arrival charts, which are multicolored, single sheets, approximately 15 × 20 in, that provide pilots with a 50-nm-radius depiction, at a scale of 1:500,000 (sectional), of terrain data and low-altitude airways for navigation enroute to terminal transition to or from selected airports. Data required for navigation under IFR are included. Selected airports, special-use airspace, tint contours, and the highest spot or obstruction elevations are also shown.

Jeppesen produces chart series that cover the world. Planning, low- and high-altitude enroute, and area charts are available for sale on a one-time basis or annual subscription. Jeppesen publishes a product catalog; see addresses and telephone numbers in App. B.

Chart series that cover Canada and selected other areas, are produced by the Canada Map Office. Planning, low- and high-altitude enroute, and area charts are available for sale on a one-time basis or annual subscription. The Canadian Map Office also publishes a catalog; see addresses telephone numbers, and Internet sites in App. B.

Canada publishes and distributes several enroute chart series. The enroute low- and high-altitude charts provide aeronautical information for instrument navigation in the low and high airway structures. Charts cover Canada, Greenland, and portions of the contiguous United States and Alaska. Charts are also available for the Azores, Bermuda, and Iceland. These charts are revised every 56 days. Charts are printed back to back on eight sheets. Terminology and symbols are similar to those used on United States instrument charts.

Enroute low-altitude charts, labeled LE, depict aeronautical radio navigation information, special-use airspace, and selected airports. Stations with communications are tabulated in alphabetical order. Like United States enroute low altitude charts, these charts provide coverage from the surface up to, but not including, 18,000 ft MSL. These charts are supplemented by a series of terminal area charts that depict aeronautical radio navigation information in congested areas at a larger scale. Enroute high-altitude charts, labeled HE, depict aeronautical radio navigation information, selected airports, and special-use airspace, for flights at and above 18,000 ft MSL.

6

Departure procedures (DP) and standard terminal arrival route (STARs) charts

Standard instrument departure (SID) procedure charts were introduced in 1961, with standard terminal arrival route (STAR) procedure charts in 1967. Profile descent charts were published in 1976. In 1999 the FAA combined the traditional standard instrument departure (SID) and textual IFR departure procedures into a single procedure called an *instrument departure procedure* (DP). DPs are published in either text or chart form.

Clearance delivery procedures have been simplified at many airports with establishment of instrument departure procedures and standard terminal arrival routes. The purpose of these procedures is to reduce controller and pilot workload. DPs facilitate transition between takeoff and the enroute phase of flight; STARs aid the transition between the enroute phase and the approach segment. They provide graphic or graphic and textual descriptions of departure and arrival clearances. This reduces controller and pilot workload by allowing the controller to assign an often complex route including altitude restrictions in a single phrase, the name of the DP or STAR. The pilot then has a printed copy of the instruction. Filing available DPs and STARs, is recommended. If, for whatever reason, a pilot does not have DPs or STARs the contraction DSNO (DPs STARs no) should be entered in the remarks section of the flight plan.

DPs and STARs are published along with instrument approach procedure (IAP) charts in one of 24 top-bound or looseleaf NACO terminal procedure publication (TPP) volumes,

5 3/8 × 8 1/4 in, covering the contiguous United States, Puerto Rico, and the Virgin Islands. [Volume Southeast 3 (SE-3) includes Puerto Rice and the Virgin Islands.] Alaskan DPs and STARs are available in the Alaska terminal procedures publication, which is similar to the TPP—this publication also contains military terminal procedures. For the Pacific area, DPs and STARs are in the *Pacific Chart Supplement*, discussed in Chap. 8. Charts are generally not to scale. These publications are designed for civil and military use. DPs and STARs, along with charted visual flight procedures (CVFPs), are listed in the index of the TPP by city and airport name. Figure 6-1 contains United States terminal procedures publication coverage. Features include:

- Inoperative components table
- Explanation of terms/landing minima format
- Index of terminal charts and minimums
- IFR takeoff and departure procedures, including DPs
- Rate of climb table
- IFR alternate minimums
- General information and abbreviations
- Chart legends
- Frequency pairing
- Radar minimums
- STARs
- Terminal charts
- Rate of descent table

Volumes are published every 56 days, with a 28-day midcycle change notice volume. Changes that occur between the 28-day cycles are published in the form of FDC NOTAMs and incorporated into the *Notice to Airmen* (NOTAM) publication.

This chapter discusses the following charts and sections in the terminal procedures publications:

- Instrument departure procedure charts (DPs)
- Rate of climb table

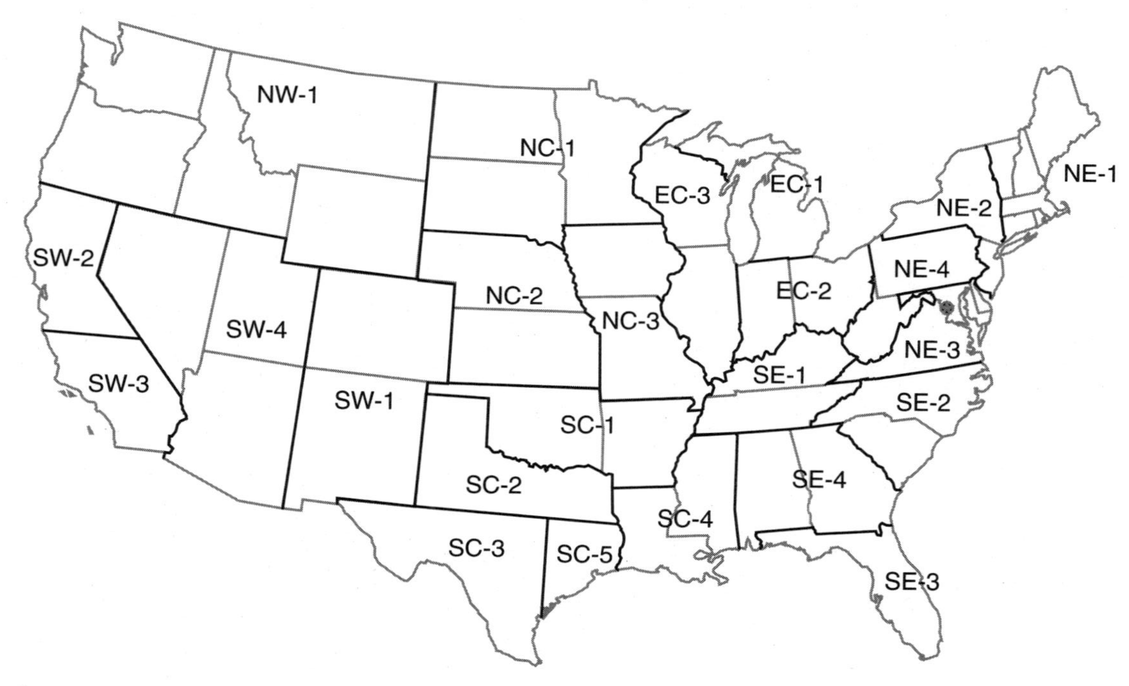

6-1 *Twenty-four terminal procedures publications contain DPs, STARs, and CVFPs for the contiguous United States.*

- Standard terminal arrival charts (STARs)
- Charted visual flight procedures (CVFPs)

Certain STARs, formally known as a *profile descent,* allow a pilot an uninterrupted descent, except for level flight required for speed adjustments, from the enroute structure to the point of glide slope intercept, or a specific minimum altitude. Charted visual flight procedures have been created to move air traffic safely and expeditiously during periods of relatively good weather. Charted visual flight procedures supplement conventional visual approach procedures by providing specific routes and altitudes, at times for noise abatement purposes.

Jeppesen DPs and STARs use a different chart indexing system from the NACO terminal procedures publication. Charts are indexed alphabetically by city name within each state. Each airport listed under the city name is given an index number, enclosed in an oval, centered in the chart top margin. Unlike the TPP, which places STARs in a separate section of the volume, Jeppesen DP and STAR charts are filed along with the instrument approach procedure charts for the airports they serve. Jeppesen provides latitude and longitude coordinates, to the tenth of a minute, rather than hundredths used on NACO charts. This is helpful because most coordinate navigation systems deal with latitude and longitude to the nearest tenth of a minute. Otherwise, charts are similar in content and format to NACO products. Jeppesen, like the TPP, includes a rate of climb table to convert ground speed and rate of climb into climb gradient in feet per nautical mile. Jeppesen's gradient to rate table is more elaborate; therefore, it is easier to use because less interpolation is required.

The bottom margin on Jeppesen charts provides a brief description of the last change—which now has been added to NACO charts; small arrows in the text indicate changes from the previous issuance. The change description and arrows indicate changes for pilots who regularly use the procedure, which can be extremely helpful.

It's important to note effective dates and not use any chart or procedure before it becomes effective, unless specified otherwise in NOTAMs. Pilots have filed DPs or STARs that had not yet become effective.

Symbols

Symbols used on DPs and STARs are similar to those on other instrument charts. Symbols unique to DPs and STARs are contained in Fig. 6-2. Airport symbols are standard. Radials and bearings are magnetic, mileages nautical, and altitude and elevations are feet above mean sea level (MSL).

In addition to those NAVAID symbols already mentioned, Fig. 6-2 contains several additional depictions. The locations of localizer (LOC) and localizer with DME (LOC/DME) antennas are shown as necessary. Seventy-five MHz marker beacons, or fan markers, are also depicted. Marker beacons may be located on an instrument landing system (ILS) at the outer

NAVIGATIONAL AIDS Localizer (LOC)		ROUTES MEA/MOCA Departure/Arrival Route Mileage	4500 *3500 270° (65)
Localizer/DME (LOC/DME)		Transition Route	
		Radial Line & Value	R-270
Marker Beacon (OM, MM, IM)		Lost Communications Track	
Marker Beacon w/Compass Locator (LOM, LMM, LIM)		ALTITUDES Mandatory Altitude	5500
Localizer Course		Minimum Altitude	2300
		Maximum Altitude	4800
Simplified Directional Facility Course (SDF)		Recommended Altitude	2200
WAYPOINT	PRAYS N38°58.30' W89°51.50' 112.7 CAP 187.1° 56.2 590	HOLDING PATTERN	(IAS)
DISTANCE NOT TO SCALE		TAKEOFF/ DEPARTURE PROCEDURE	T

6-2 *SID and STAR symbols are similar to those used on other instrument charts.*

marker (OM), middle marker (MM), or inner marker (IM). At many locations a compass locator or NDB is collocated with a marker beacon. A compass locator may be located at any one of the three marker beacon locations. As necessary, localizer course—with the feathered right or blue sector—and simplified directional facility courses (SDF) are shown.

NAVAID boxes do not indicate FSS frequencies, but primary NAVAIDs indicate name, frequency, frequency protection range (SSV), identification, and RNAV coordinate information. RNAV coordinate information is omitted from secondary NAVAIDs. Figure 6-2 illustrates a waypoint (W/P) identification box. PRAYS is the name of the W/P. The next line contains its latitude/longitude coordinates, the bottom line the W/P's VORTAC RNAV description. At the base of the box is the reference facility elevation (590).

Refer to Fig. 6-2. Departure and arrival routes are depicted by a heavy black line with arrow, transition routes a medium black line, and radials with a thin black line. Lost communication tracks are shown as dotted lines. As necessary, MEA and MOCA altitudes are depicted. Mandatory altitudes are indicated by lines above and below the altitude, minimum altitude by a line below the altitude, maximum altitude by a line above the altitude; the omission of lines above and below indicates recommended altitudes.

As necessary, holding patterns, along with maximum holding airspeeds (IAS), are depicted. When takeoff minimums are not standard or a published departure procedure is available, this will be noted with the white letter T in a black triangle. Air traffic clearance instructions appear on these charts.

Charted visual flight procedure (CVFP) charts depict prominent landmarks, courses, recommended altitudes to specific runways, and NAVAID information for supplemental navigational guidance. Procedure tracks are indicated by a dashed heavy black line with a series of arrowheads. Routes may contain approximate headings. If landmarks used for navigation will not be visible at night, the approach will be annotated: PROCEDURE NOT AUTHORIZED AT NIGHT. Reference features and obstacles are designated, sometimes along with their height. CVFPs usually begin within 15 miles of the airport, with published weather minimums based on minimum vectoring altitudes; they are not instrument approaches, and do not contain missed approach procedures.

Instrument departure procedures (DPs)

Departure procedures that are necessary for basic obstacle clearance during departure, and are designated as the primary obstacle avoidance departure for a specific runway, will be listed by airport in "IFR Take-Off Minimums and Departure Procedures," Sec. C of the terminal procedures publication (TPP). If the basic obstacle avoidance DP is textual only, it will be described in Sec. C of the TPP. If the DP is a charted graphic procedure, the procedure will be named, and the name listed by airport in "Index of Terminal Charts and Minimums," Sec. B of the TPP.

The use of a DP requires the pilot to have at least the textual description of the procedure. If the DP contains the departure control frequency, it might be omitted from the clearance. Graphical DP features include:

- Radio aids to navigation
- Communication frequencies
- Reporting points
- Airways and mileages
- Holding patterns
- Special-use airspace
- Airports
- IFR takeoff minimums and departure procedures table
- Airport sketch
- Departure route description
- Geographic positions of NAVAIDs and reporting points
- Mileage breakdown points
- Changeover points
- Computer codes for filing flight plans
- Data for coordinate navigation systems

Departure procedures assist pilots in avoiding obstacles during climbout. Obstacle clearance is based on the aircraft climbing at the rate of at least 200 ft per nautical mile (fpnm), crossing the end of the runway at least 35 ft AGL, and climbing to 400 ft AGL before turning, unless otherwise specified in the procedure. Climb gradients are specified when required for obstacle clearance. Crossing restrictions might be established for traffic separation or obstacle clearance. When no gradient is specified, the pilot is expected to climb at a rate of at least 200 ft per nautical mile to the MEA. To assist pilots in determining required climb rate, the TPP contains a rate of climb table on page D1. The rate of climb table is illustrated in Fig. 6-3. The table converts rate of climb in feet per minute to rate of climb in feet per nautical mile, based on the aircraft's ground speed. Note that this is ground

REQUIRED GRADIENT RATE (ft. per NM)	GROUND SPEED (KNOTS)						
	30	60	80	90	100	120	140
200	100	200	267	300	333	400	467
250	125	250	333	375	417	500	583
300	150	300	400	450	500	600	700
350	175	350	467	525	583	700	816
400	200	400	533	600	667	800	933
450	225	450	600	675	750	900	1050
500	250	500	667	750	833	1000	1167
550	275	550	733	825	917	1100	1283
600	300	600	800	900	1000	1200	1400
650	325	650	867	975	1083	1300	1516
700	350	700	933	1050	1167	1400	1633

REQUIRED GRADIENT RATE (ft. per NM)	GROUND SPEED (KNOTS)					
	150	180	210	240	270	300
200	500	600	700	800	900	1000
250	625	750	875	1000	1125	1250
300	750	900	1050	1200	1350	1500
350	875	1050	1225	1400	1575	1750
400	1000	1200	1400	1600	1700	2000
450	1125	1350	1575	1800	2025	2250
500	1250	1500	1750	2000	2250	2500
550	1375	1650	1925	2200	2475	2750
600	1500	1800	2100	2400	2700	3000
650	1625	1950	2275	2600	2925	3250
700	1750	2100	2450	2800	3150	3500

6-3 *The TPP provides a table to convert climb rate in feet per minute to rate of climb in feet per nautical mile, based on aircraft's ground speed.*

speed, not airspeed; therefore, winds during the climb must be taken into account.

For example, at Bakersfield, Calif., the Wring Three Departure requires a minimum climb gradient of 345 fpnm to 5400 ft MSL. Bakersfield has an elevation of approximately 500 ft. Therefore, the aircraft must climb at 345 fpnm for 4900 ft in a distance of 15 nm. The 1967 Cessna 172 manual advertises a sea level climb rate of 645 fpm; at 5000 ft, 435 fpm, at an indicated airspeed of 70 knots. Assume a ground speed of 70 knots—which cannot necessarily be presumed in the real world—for interpolating a rate of climb. From Fig. 6-3 we see that an approximately rate of climb of 410 fpm would meet this requirement. At standard conditions, the 1967 Cessna 172 should be able to comply with this DP's climb gradient. It's extremely important for pilots to carefully consider ground speeds and aircraft performance before accepting a DP with an increased climb gradient.

Departure procedures can be divided into two general categories: pilot navigation departures and radar vector departures.

Pilot navigation departure procedures

With pilot navigation departure procedures, the pilot assumes primary responsibility for navigation. Some pilot navigation DPs might contain vector instructions, until the controller issues a clearance to resume normal navigation on the filed or assigned route, or DP procedure.

Figure 6-4 shows the Livermore Municipal Airport (LVK), Livermore, California, LIVERMORE ONE DEPARTURE and BYRON ONE DEPARTURE. Note that both departures are PILOT NAV and their computer codes are respectively LIVR1.ALTAM and BYRON1.BYRON. (More about the proper use of computer codes in the following section.) After the computer code is the number 01249. This is the Julian date when the chart was issued. (The Julian calendar numbers the days of the year consecutively from 001, which is January 1; the year precedes the three-digit day group. The Julian date in the example is 01249; this procedure was published on September 6, 2001.) The number SL-6075 is the FAA's chart identification number.

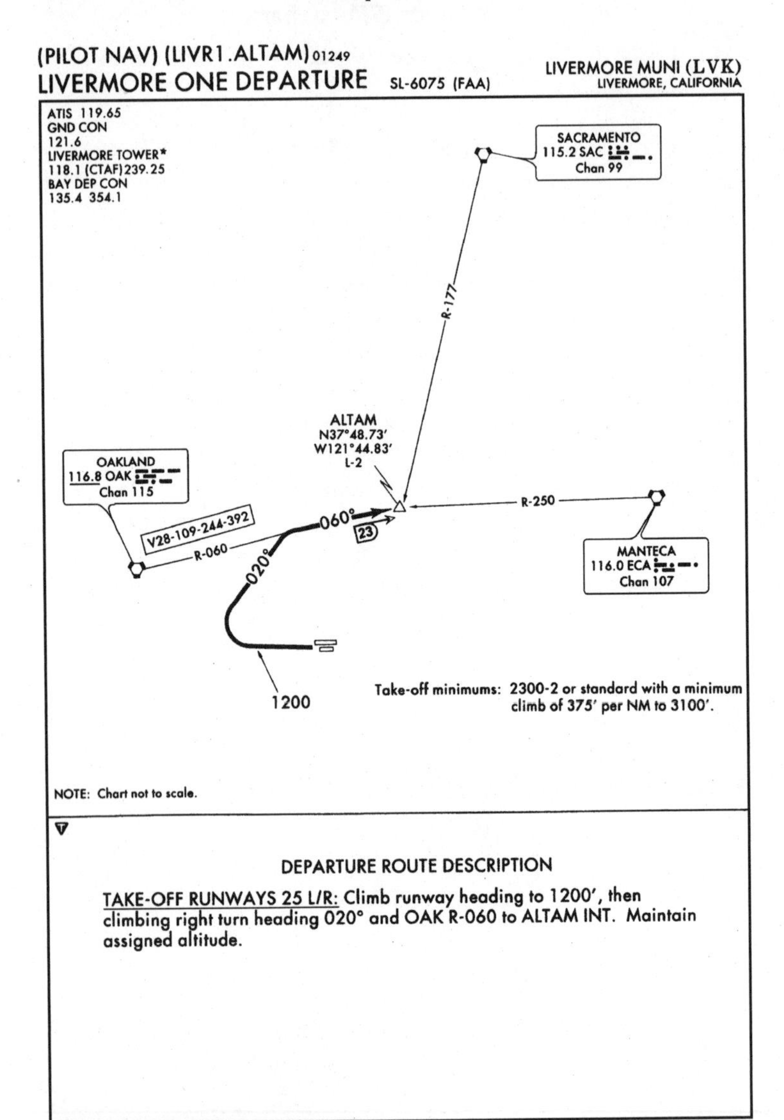

6-4 *With pilot navigation departure procedures, the pilot assumes primary responsibility for navigation.*

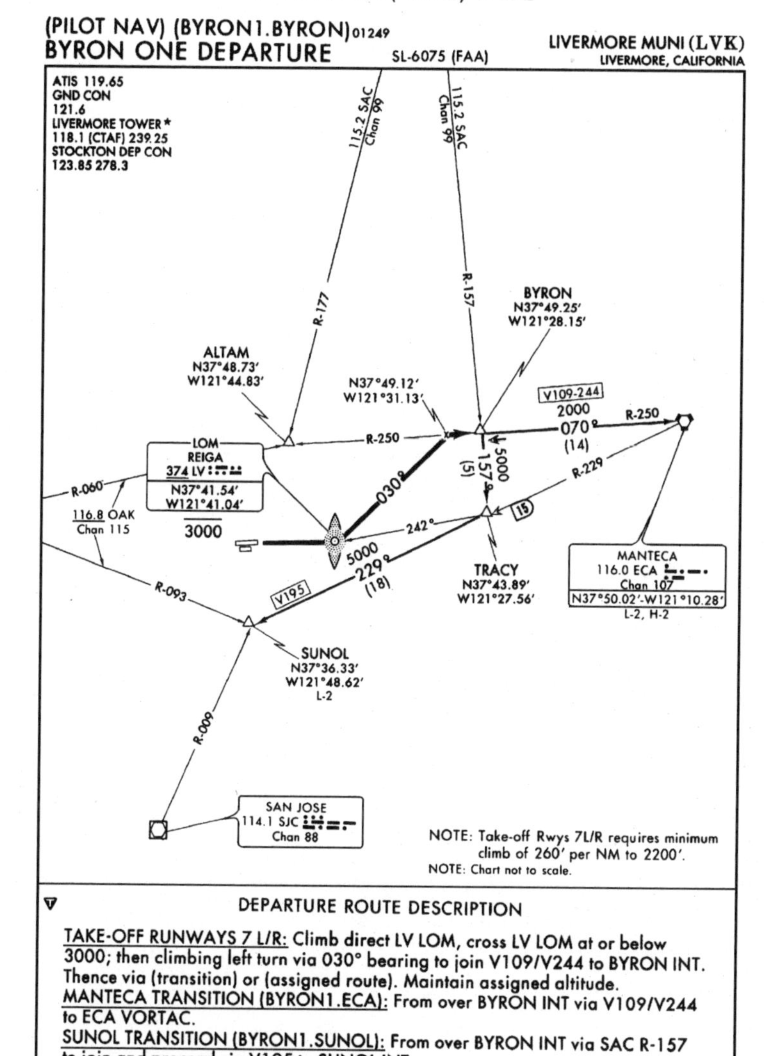

6-4 *(Continued).*

Appropriate communication frequencies are provided in the upper left portion of the plan view. Note the communication frequencies 239.25 and 354.1. These are UHF communication frequencies for military aircraft. This accounts for those occasions where we may only hear the controller's part of the conversation when they are dealing with military aircraft.

Case Study

As a brand new private pilot, I flew our aero club Cessna 150 from our base at Woodbridge to Weathersfield in East Suffolk, England. I picked up an Air Force officer who was a flight instructor, but not an Air Force pilot. The weather turned bad as we prepared for takeoff. The instructor pilot changed our flight plan from VFR to IFR and obtained a clearance to Woodbridge. We were assigned a departure frequency of 285.4 MHz. I watched in amazement as this officer attempted to dial 285.4 into our 90-channel VHF radio! I informed him it wasn't there and we obtained a VHF frequency for the trip.

It cannot be overemphasized how important it is for pilots to read and understand all of the notes and information on instrument charts. Refer to the LIVERMORE ONE departure in Fig. 6-4. One note states: "Chart not to scale." The other note informs the user that takeoff minimums are 2300-2 or standard with a minimum climb of 375 fpnm to 3100 ft. What does this mean and how does it affect the pilot? In this case there are mountains along the departure course. However, with a ceiling of 2300 ft and 2 miles visibility the pilot can visually avoid terrain with the standard climb gradient. Otherwise, the pilot is required to be able to climb at a rate of 375 fpnm to 3100 ft to ensure obstacle clearance.

Do takeoff minimums apply to 14 CFR Part 91 operations? A check of the regulations reveals they do not. Does this mean it's safe for pilots to take off in zero-zero conditions? Absolutely not! Nor does this exemption relieve the pilot from complying with other parts of the regulation, such as: "An altitude allowing, if a power unit fails, an emergency landing without undue hazard to persons or property on the surface." And let's not forget careless or reckless operations.

(Some instructors teach zero-zero takeoffs. Although not prohibited, this is certainly not the behavior we are trying to instill in our instrument students.)

Let's move on to the departure route description. The first thing to note is that the LIVERMORE ONE applies only to takeoff from runways 25 left and 25 right. The textual description instructs pilots to climb on runway heading to 1200 ft, then make a climbing right turn to a heading of 020° and the OAK 060° radial to the ALTAM intersection. (The OAK 060° radial defines V28-109-244-392 between the OAK VOR and ALTAM.) This route is depicted graphically in the plan view by the heavy black line. The plan view also provides applicable navigational frequencies.

What would be the minimum aircraft navigational equipment to comply with this departure? A single VOR receive would allow the pilot to fly this departure. What if the Sacramento (SAC) VOR were out of service? The SAC 177° radial is required to establish the ALTAM intersection, the exit fix. Therefore, a pilot with VOR only could not accept this departure. Are there other means to identify ALTAM? Yes. DME from OAK or an approved area navigation system—note the coordinates for ALTAM. What about the Manteca (ECA) 250° radial? Since there is less than a 30° divergence between the OAK and ECA radials, this would not be an appropriate means to establish ALTAM.

Below the intersection name is the label L-2. This informs the pilot that ALTAM can be located on enroute low-altitude chart L-2.

Case Study

Recently the SAC VOR was out of service. Someone issued a NOTAM that the LIVERMORE ONE departure was unusable and this information was placed on the ATIS. Subsequently, someone realized that ALTAM could be established with DME or GPS and the ATIS broadcast was revised.

In spite of this instance, pilots cannot expect ATC to issued NOTAMs of this type or know if the aircraft has the navigational capability to comply with a specific clearance. This responsibility rests solely with the pilot.

Finally, note the white T in the black triangle in the departure route description. This indicates takeoff minimums are not standard or a published departure procedure is available, which can be found in the takeoff minimums and obstacle departure procedures section (Sec. C) of the TPP. As it turns out these are the same instructions, in text form only, for the two Livermore DPs. Therefore, a pilot having only the text instruction could accept either of these departure procedures.

Refer to the BYRON ONE departure in Fig. 6-4. The chart is similar in format to the LIVERMORE ONE departure. This procedure applies to runways 7 left and 7 right only. The departure instructs the pilot to climb direct to the Livermore outer compass locator (LV LOM), cross the LOM at or below 3000 ft. (This restriction is for ATC separation requirements, rather than obstacle avoidance. ATC separation is also a function of many DPs.) Over the LOM the pilot is instructed to initiate a climbing left turn via the 030° bearing from the LOM to join V109/V244 (ECA R-250) to the BYRON intersection. Since navigation is based on an NDB, the aircraft must be equipped with an operational automatic direction finder (ADF) or an approved RNAV substitute.

The end of the BYRON ONE, or the exit fix, is the BYRON intersection. This is depicted as the thick black line in the plan view. Unlike the LIVERMORE ONE, the BYRON ONE departure has two transitions—Manteca and SUNOL. The Manteca transition (BYRON1.ECA) takes the pilot from over BYRON to the ECA VOR; the SUNOL transition (BYRON1. SUNOL) takes the pilot from over BYRON, via the SAC 157° radial to the TRACY intersection and then the ECA 229° radial to the SUNOL intersection. These transitions are depicted in the plan view by the medium black lines. The pilot would then change to the enroute structure from either ECA or SUNOL.

Below the ECA NAVAID box are the labels L-2 and H-2. This informs the pilot that ECA can be found on enroute low-altitude chart L-2 and enroute high-altitude chart H-2.

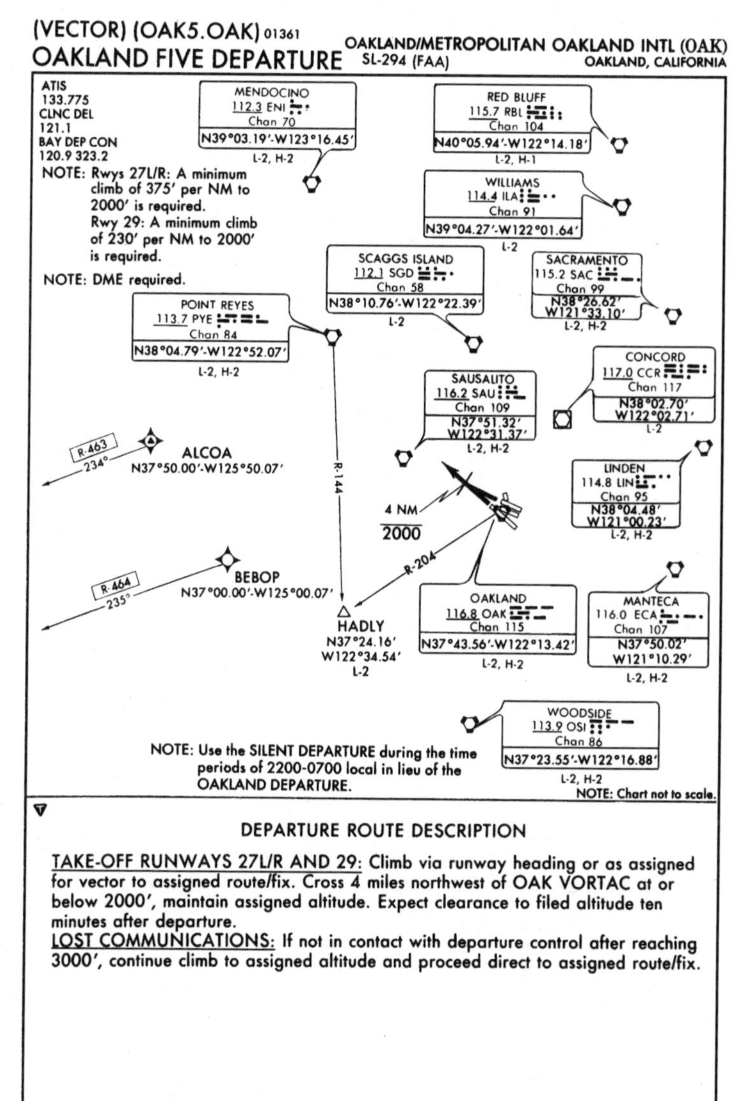

6-5 *Air traffic control assumes primary responsibility for navigation with vector departure procedures.*

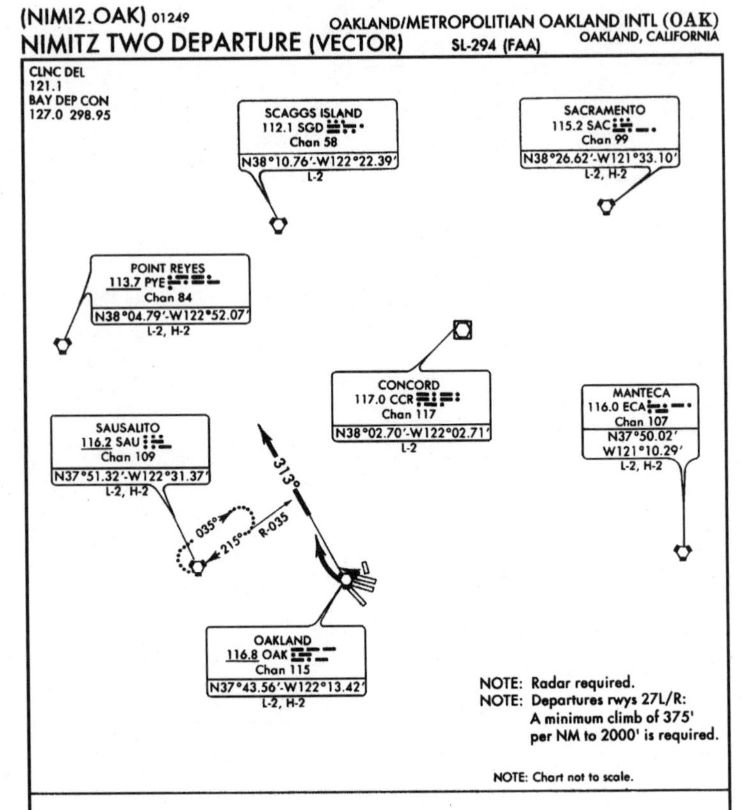

DEPARTURE ROUTE DESCRIPTION

TAKE-OFF RUNWAYS 27L/R and RUNWAY 29: Turn right, intercept and proceed via OAK R-313 for vectors to assigned fix/route. Expect clearance to filed altitude 10 minutes after departure.
LOST COMMUNICATIONS: If not in contact with departure after reaching 3000′, proceed direct to SAU VORTAC and hold on SAU R-035. Climb in holding pattern to assigned altitude, then proceed to assigned fix/route.

NIMITZ TWO DEPARTURE (VECTOR) OAKLAND, CALIFORNIA
(NIMI2.OAK) 01249 OAKLAND/METROPOLITAN OAKLAND INTL (OAK)

6-5 *(Continued).*

Vector departure procedures

Vector DPs are established where ATC assumes primary radar navigational guidance to the filed or assigned route. Again, should a T appear in the route description, it would indicate nonstandard takeoff minimums, which can be found in the IFR takeoff minimums and departure procedures section C of the TPP.

Figure 6-5 contains two examples of vector DPs. The Oakland/Metropolitan Airport (OAK), Oakland, Calif., OAKLAND FIVE DEPARTURE and NIMITZ TWO DEPARTURE. As in pilot navigation, vector DPs most often will have a computer code (OAK5.OAK and NIMI2.OAK). Both of these departures have only limited communication data in the upper left corner of the plan view. This is because this airport has a separate airport diagram which contains all the appropriate frequencies. (The airport diagram is discussed in Chap. 7.)

Let's begin the discussion with the OAKLAND FIVE departure. Here again, careful attention to notes is required. This procedure requires that aircraft be equipped with DME, or an approved RNAV substitute. Notes also indicate that this DP is not available between the hours of 2200 and 0700 local. During this time pilots are instructed to use the SILENT DEPARTURE—for noise abatement. This procedure is available only for departures from runways 27 left, 27 right, and 29. The pilot is instructed to climb via runway heading, or as assigned, for radar vectors to the assigned route or fix; cross 4 miles northwest of the OAK VORTAC at or below 2000 ft. Why? Like the altitude restriction on the BYRON departure, this is to provide traffic separation. Failure to note and comply with these restrictions could result in loss of aircraft separation, and the possibility of ATC filing a pilot deviation.

The departure route description contains lost communications procedures. If communications have not been established with departure control after reaching 3000 ft, the pilot is to continue climb to assigned altitude and proceed direct to the assigned route or fix. Normally, one of the plan view fixes will be part of the route clearance.

The NIMITZ TWO departure also contains lost communications procedures. This includes an instruction to hold at the Sausalito VORTAC. Note that the holding pattern is depicted with a doted line, indicating that the hold is part of the lost communications procedure.

The computer code for the NIMITZ TWO is NIMI2.OAK. Most vector DPs use this format. Pilots filing this procedure will normally file the departure airport, departure number, exit fix—for example: OAK NIMI2 OAK. Pilots are expected to begin the enroute phase at one of the fixes depicted on the plan view. Pilots should choose an appropriate fix for the altitude structure they plan to fly, low or high. For example: "...OAK NIMI2 OAK SAC...." (Pilots filing through DUAT or other computer systems will use a space in place of the dots in the computer code. The reason for this is a long and ugly story. The bottom line: Use a space between route elements when filing through DUATs.)

Some vector DPs do not have computer codes—for example, the Bakersfield, Calif., MEADOWS ONE departure (vector). It is essentially a radar vector to an enroute NAVAID. Since there is no computer code, it cannot be entered in the route section of a flight plan. Pilots, however, can expected to be issued the departure, and, if they wish, could put the DP in the remarks sections of the flight plan.

Using departure procedures

Information on traffic delays is an essential part of a pilot's preflight preparation. In fact, the regulations state: "For a flight under IFR...any known traffic delays of which the pilot in command has been advised by ATC...." This is most often accomplished through the FAA's Notice to Airmen system. (More specifics about NOTAMs in Chap. 8.) With this scheme, known as *flow control,* a pilot basically has two options. First, as most of the air carriers do, call for clearance in the ramp area and obtain an estimated release time. Second, taxi out to the runup area, complete the pretakeoff checklist, and await the release time. (A pilot may wish to monitor and copy departure ATIS, then contact clearance delivery. Departure delays and flow control delays are usually part of the departure ATIS. Many pilots prefer doing this before starting the engine, especially when delays are possible; a hand-held transceiver is ideal for this task.)

Case Study

It was overcast in the Bay Area because of coastal stratus. We had filed from Livermore to San Jose International. (Actually that is Mineta San Jose International now; we sure have a propensity to name things.) We were advised of a 30-minute delay. When delays are more than about 10 to 15 minutes, I prefer to shut down the airplane and monitor ground or tower, as appropriate. Shortly before our release time we restart the engine. Exactly, exactly 30 minutes later we were released. We still received some delay vectors, but the men and women at Bay Approach, bless 'em, squeezed our Cessna 172 into the jet flow at San Jose.

A pilot should review available DPs for the departure airport during preflight planning. Note that some DPs, for some airports, are used only during certain traffic flow periods. Some DPs apply to all runways, others don't. For example, in the San Francisco Bay area, traffic is divided into two basic flow plans. Normal traffic is to the west, but during winter storms, because of strong southerly winds, traffic lands and departs to the southeast. The pilot must note which runways a DP serves. A pilot can normally determine departure runway for present and forecast wind direction and speed. A flight plan filed with a DP that is not in effect will normally be rejected.

A pilot must also determine during the preflight planning if the DP applies to the type of operation; some DPs apply to jets only, others exclude jet aircraft. The DP exit fix or transition must connect with the enroute structure to be flown. Pilots who fail to consider these factors might be assigned departure procedures that they did not file, or encounter delays while ATC attempts to find and assign a suitable route. When in an unfamiliar area, local pilots or the FSS can recommend appropriate departure routes.

The pilot must also ensure that the aircraft is equipped to comply with the procedure. Some DPs require the use of special facilities, such as radar, ADF, or DME. Pilots should not accept DPs with which their aircraft performance or equipment cannot comply.

Case Study

> *Departing San Francisco International in a Cessna 172, I was assigned, by clearance delivery, a DP that required the use of DME to establish a turning point on the departure. The aircraft was not equipped with DME, and I informed the controller; however, the controller was able to provide me with a VOR radial, not on the DP chart, to establish the fix, and I was able to accept the departure.*

Selecting a suitable DP allows the pilot to study the procedure, becoming familiar with NAVAIDs, route, and altitudes. This will assure that, once assigned by ATC, the procedure can be flown, especially if the DP has an increased climb gradient or special NAVAID requirement.

Flight plans with incorrect DP or transition codes are rejected. Pilots filing DPs must use the correct departure codes and exit fix or transition; the filed airway structure must begin at the exit or transition fix. For example, a pilot plans to depart Oakland and proceed via V27 northbound. (V27 is shown in Fig. 5-10). When you are filing through an FSS or DUAT, the departure procedure, like the route, must connect. The following is a correct route:
OAK NIMI2 OAK PYE V27…

OAK is the departure airport, NIMI2 the DP with OAK as the exit fix. At Point Reyes (PYE) the pilot plans to pick up V27. This is important since an incorrect route will result in the flight plan being rejected. Below are examples of incorrect routes:

OAK NIMI2 PYE V27…

The exit fix is missing.

OAK NIMI2 OAK V27…

OAK is not a fix on V27.

One final thought about departures. Extensive use, in certain parts of the country, is made of IFR to VFR on top. These clearances often include a DP. In effect an IFR to VFR on top is a clearances to nowhere! Let's take, for example, "...Cleared to the XYZ VOR via the Headsup Two departure XYZ transition, climb and maintain 3000, if not on top at 3000, maintain 3000 and advise...." After departure we lose radio communications, arrive at XYZ at 3000 and we're still in the clouds! Lost communications procedures don't cover this possibility. A prudent pilot would be very cautious accepting such a clearance unless tops were known. If not, it might be wise to file to a destination, then cancel once in the clear on top.

Standard terminal arrival routes (STARs)

A standard terminal arrival route is an ATC-coded IFR arrival route established to expedite IFR aircraft destined for certain airports. In effect STARs are the reverse of graphic DPs. Flight management system procedures (FMSPs) serve the same purpose, but are only used by aircraft equipped with flight management systems (FMSs).

Flight Management System

This is a computer system that uses a database to allow routes to be preprogrammed and placed into the computer. The system is constantly updated with respect to position accuracy by reference to conventional navigation aids. The computer program automatically selects the most appropriate aids during the update cycle. Aircraft so equipped use the aircraft suffix codes /E or /F.

The purpose of a STAR is to simplify clearance delivery procedures and facilitate transition between enroute and instrument approach procedures. STARs provide graphic and textual descriptions of preplanned IFR air traffic control arrival procedures. They reduce pilot and controller workload and communications, and minimize potential errors in delivery and receipt of clearances. As with DPs, pilots are requested to use associated STAR codes when filing flight plans. Features include:

- Geographic positions for NAVAIDs and reporting points
- Radio aids to navigation

- Communication frequencies
- Routes
- Assigned altitudes
- Special-use airspace
- Mileages
- Mileage breakdown points
- Airports
- Reporting points/fixes
- Holding patterns
- Minimum enroute altitudes
- Minimum obstruction clearance altitudes
- Computer codes for filing flight plans
- Changeover points
- Data for area navigation systems

STARs and FMSPs may contain mandatory speeds and crossing altitudes. Others may have planning information to inform the pilot what clearances or restrictions to anticipate. Any "expect" altitude or speed is not considered a crossing restriction until verbally issued by the controller; published "expect" altitude or speed information is for planning purposes only. It is not to be used in the event of lost communications unless ATC has specifically advised the pilot to expect these restrictions as part of a further clearance.

Pilots assigned these procedures must maintain the last assigned altitude until they receive authorization to descend, so as to comply with all published or issued restrictions. This authorization is contained in the phraseology "...descend via...." A "...descend via..." clearance authorizes the pilot to vertically and laterally navigate, in accordance with the depicted procedure, to meet published restrictions. Vertical navigation is at "pilot's discretion"; however, adherence to published altitude crossing restrictions and speeds is mandatory unless otherwise cleared.

Pilots destined to airports with published STARs may be issued a clearance containing a STAR whenever deemed appropriate by ATC. Like DPs, the acceptance of a STAR requires the pilot to have at least the textual description of the procedure.

Figure 6-6 contains the Oakland, Calif., MANTECA ONE ARRIVAL (ECA.ECA1) and the MADWIN FOUR ARRIVAL (ECA.MADNR). Communications information is contained in the upper right-hand corner of the charts. Transitions and route description are displayed in the upper left corner, with the plan view oriented to north. Refer to the MANTECA ONE and MADWIN FOUR arrivals in Fig. 6-6. During preflight planning, a pilot's first task is to select a STAR that has a transition or entry fix on the enroute portion of the flight—either low or high altitude. The two examples in Fig. 6-6 have Coaldale, Mina, and Mustang as transition fixes and Manteca as the entry fix. Therefore, we could enter the arrival at any one of the transition fixes, or at Manteca the entry fix to both procedures.

So what is the difference between the two procedures? The MANTECA ONE terminates at the UPEND intersection; the MADWIN FOUR terminates at the SUNOL intersection. (We know this because these intersection are at the end of the thick black line that represents the arrival route.) Like DPs, STARs may serve only one runway or one set of runways. The MANTECA ONE arrival is used when Oakland is landing on runways 11 or 9 right; the MADWIN FOUR arrival is used when Oakland is landing on runways 29, or 27 left or 27 right. A review of Oakland's instrument approach procedures reveals that virtually all the approaches use COMMO or SUNOL intersections as intermediate or final approach fixes.

Note that both procedures in Fig. 6-6 contain VERTICAL NAVIGATION PLANNING INFORMATION. (This information was formerly contained in profile descents.) Let's say we were inbound to Oakland via J80. From forecast surface winds at Oakland, we determine that the most favored runway for our arrival is 9 right. Therefore, we have filed ...J80 OAL ECA1 OAK. The procedure description informs us that from over the OAL VORTAC we are to proceed via the OAL R-250 and the ECA R-068 to the ECA VORTAC; thence...from over ECA via the ECA R-229 to LOCKE intersection, then via SGD R-107 to UPEND intersection; expect radar vectors to the final approach course.

Since most of the approaches to Oakland do not have the UPEND intersection as either an initial or intermediate

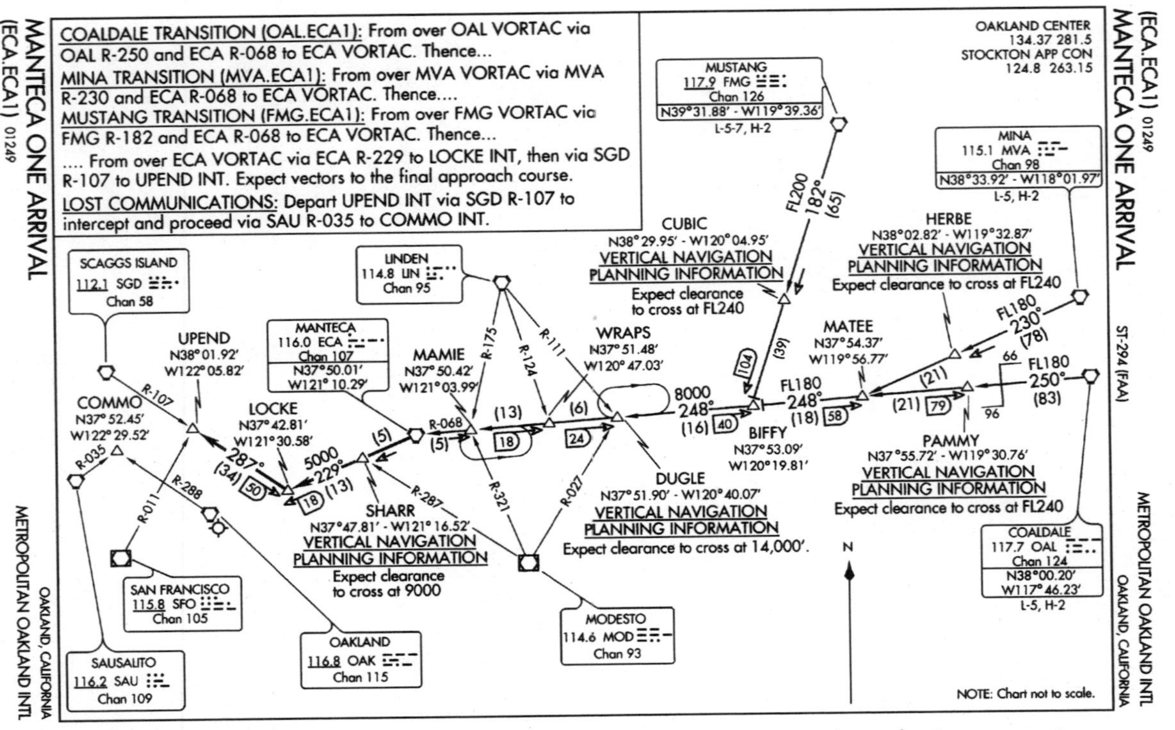

6-6 *STARs facilitate an aircraft's transition from the enroute structure to the instrument approach procedure.*

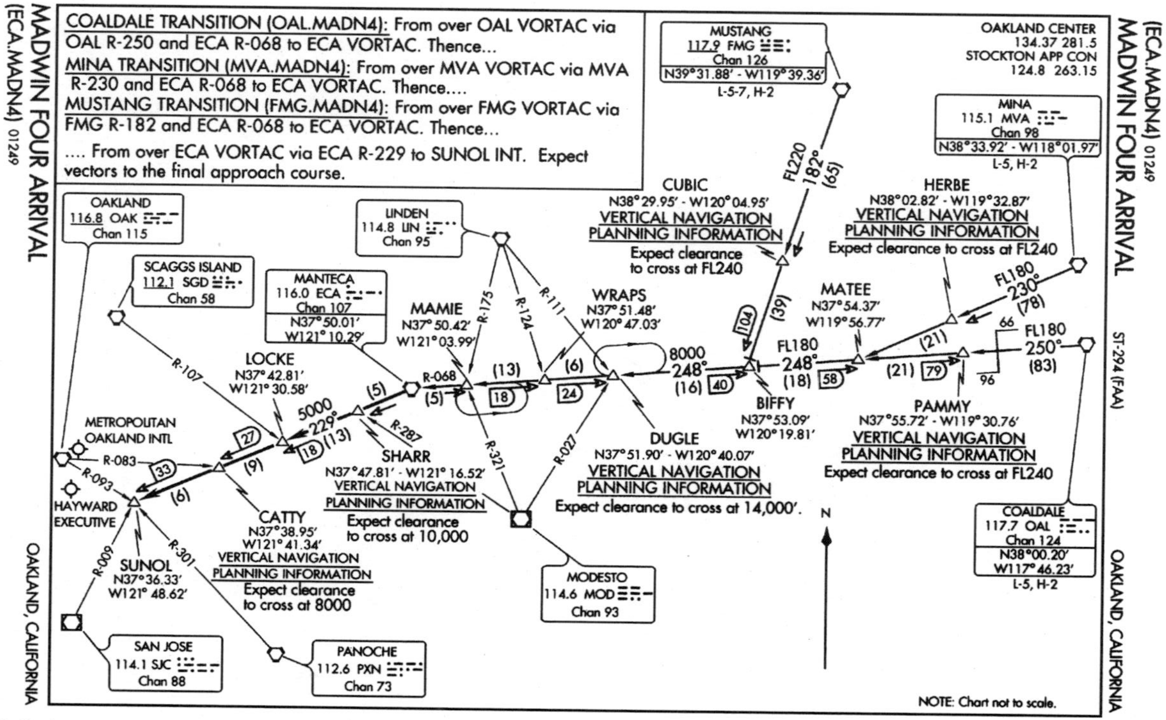

(ECA.MADN4) 01249

MADWIN FOUR ARRIVAL

ST-294 (FAA)

OAKLAND, CALIFORNIA

COALDALE TRANSITION (OAL.MADN4): From over OAL VORTAC via OAL R-250 and ECA R-068 to ECA VORTAC. Thence...

MINA TRANSITION (MVA.MADN4): From over MVA VORTAC via MVA R-230 and ECA R-068 to ECA VORTAC. Thence....

MUSTANG TRANSITION (FMG.MADN4): From over FMG VORTAC via FMG R-182 and ECA R-068 to ECA VORTAC. Thence...

.... From over ECA VORTAC via ECA R-229 to SUNOL INT. Expect vectors to the final approach course.

OAKLAND CENTER
134.37 281.5
STOCKTON APP CON
124.8 263.15

MUSTANG
117.9 FMG
Chan 126
N39°31.88' - W119°39.36'
L-5-7, H-2

MINA
115.1 MVA
Chan 98
N38°33.92' - W118°01.97'
L-5, H-2

COALDALE
117.7 OAL
Chan 124
N38°00.20'
W117° 46.23'
L-5, H-2

OAKLAND
116.8 OAK
Chan 115

LINDEN
114.8 LIN
Chan 95

SCAGGS ISLAND
112.1 SGD
Chan 58

MANTECA
116.0 ECA
Chan 107
N37° 50.01'
W121° 10.29'

MODESTO
114.6 MOD
Chan 93

SAN JOSE
114.1 SJC
Chan 88

PANOCHE
112.6 PXN
Chan 73

CUBIC
N38° 29.95' - W120° 04.95'
VERTICAL NAVIGATION
PLANNING INFORMATION
Expect clearance
to cross at FL240

HERBE
N38°02.82' - W119°32.87'
VERTICAL NAVIGATION
PLANNING INFORMATION
Expect clearance to cross at FL240

PAMMY
N37°55.72' - W119°30.76'
VERTICAL NAVIGATION
PLANNING INFORMATION
Expect clearance to cross at FL240

MATEE
N37°54.37'
W119°56.77'

BIFFY
N37°53.09'
W120°19.81'

WRAPS
N37° 51.48'
W120° 47.03'

DUGLE
N37°51.90' - W120° 40.07'
VERTICAL NAVIGATION
PLANNING INFORMATION
Expect clearance to cross at 14,000'.

MAMIE
N37° 50.42'
W121° 03.99'

SHARR
N37°47.81' - W121° 16.52'
VERTICAL NAVIGATION
PLANNING INFORMATION
Expect clearance
to cross at 10,000

LOCKE
N37° 42.81'
W121° 30.58'

CATTY
N37°38.95'
W121° 41.34'
VERTICAL NAVIGATION
PLANNING INFORMATION
Expect clearance
to cross at 8000

SUNOL
N37°36.33'
W121° 48.62'

METROPOLITAN
OAKLAND INTL

HAYWARD
EXECUTIVE

FL220
182°
(65)

FL180
230°
(78)

FL180
250°
(83)

66
96

(21)
(21)
79

FL180
248°
(18)
58

(39)
104

8000
248°
(16)
40

(6)
24

(13)
18

R-068
(5)

(5)
R-287

5000
229°
(13)
18

27
(9)

33
(6)

R-175
R-124
R-111
R-321
R-027
R-107
R-083
R-093
R-009
R-301

N

NOTE: Chart not to scale.

MADWIN FOUR ARRIVAL

(ECA.MADN4) 01249

OAKLAND, CALIFORNIA

6-6 *(COntinued)*.

approach fix, LOST COMMUNICATIONS procedures are provided. In this case, pilots are to depart the UPEND intersection via the SGD R-107 to intercept and proceed via the SAU R-035 to the COMMO intersection. At that point commence the approach. Since the SUNOL intersection is an initial approach fix, lost communications procedures for west plan runways are not necessary. A pilot who experiences lost communications would proceed via published transitions and execute the appropriate instrument approach procedure.

Notice that the Hayward Executive Airport is depicted on the MADWIN FOUR arrival. This is because this arrival can also be used for the Hayward Airport. This arrival should be listed under the Hayward Airport in "Index of Terminal Charts and Minimum," Sec. B of the TPP. So why is the Hayward airport not depicted on the MANTECA ONE arrival? Hayward does not have an approach from the west. All approaches are from the east to runway 28L or circle to land, with the SUNOL intersection as an initial approach fix.

Any time arrivals and departures are opposite direction—such as when a pilot attempts to make an approach to Hayward, which has instrument approaches only from the east, while the bay is landing east (head-on traffic)—extensive delays can be expected. Recall the Livermore Byron departure. Since Livermore has instrument approaches landing to the west only, should wind conditions require the use of the Byron departure (head-on traffic), expect extensive delays. (I've used examples in the San Francisco Bay area because they are appropriate to the procedures discussed. However, in any part the United States, or the world for that matter, when these conditions prevail, pilots should anticipate delays.)

Using standard terminal arrival routes

Destination STARs should be reviewed during a pilot's preflight planning. Note that some STARs, for some airports, will be used only during certain traffic flow periods. The pilot should note which runways and airports the STAR serves. Some STARs serve several airports and all approaches, others serve only one airport and one runway. Additionally, there are STARs for RNAV equipped aircraft only. This information is contained in the procedure name (KSINO RNAV ONE ARRIVAL). Although traffic plans and runways are normally not directly available during preflight planning and briefing, they are implied by forecast and actual wind direction and speed, and NOTAMs. A flight plan filed with an inappropriate STAR will normally be issued by the departure controller in a standard "as filed" clearance, only to be amended near the destination when necessary. Normally, when entering the airport's arrival sector, the controller will advise the pilot which procedure to expect.

Finally, an appropriate high- or low-altitude enroute structure must connect with the STAR transition or entry fix. The pilot must also ensure the aircraft is equipped to comply with the procedure: radar, ADF, DME, RNAV, and the like. Selecting a suitable STAR allows the pilot to study the procedure, becoming familiar with NAVAIDs, routes, and altitudes.

A pilot can begin a STAR at the entry fix. For example, a pilot could file via direct or connecting airways to ECA, then the ...ECA ECA1 OAK arrival; however, a major purpose of transitions is to allow ATC to line up aircraft, providing appropriate separation, for the approach to the destination. This expedites the flow of traffic and reduces delays.

Certain STARs are designed to allow the pilot to navigate to an approach segment where the instrument approach begins, others end in radar vectors to the instrument approach procedure. In the case of radio communications failure, in the first instance, when the STAR ends where the approach begins pilots would fly assigned routes and altitudes as required by the regulations. If communications were lost immediately after departure the pilot would fly the routes assigned by ATC in the original clearance. Should communications be lost during the arrival segment the pilot would fly the expected approach clearance to the destination. Last assigned or minimum altitudes would apply.

Unlike the terminal procedures publication, which places STARs in a separate section of the volume, Jeppesen DP and STAR charts are filed along with the instrument approach procedure (IAP) charts for their respective airports.

Charted visual flight procedures (CVFPs)

Charted visual flight procedures charts (CVFPs) depict prominent landmarks, courses, and recommended altitudes to specific runways. CVFPs serve one or more of the following functions:

- Environmental concerns
- Noise considerations
- Safety of flight
- Efficiency of air traffic operations

CVFPs have been developed primarily for turbojet aircraft and can be assigned only at airports with an operating control tower. Most CVFPs depict some NAVAID information for supplemental navigational guidance only. Altitudes are recommended for noise abatement, but do not prohibit the pilot from flying other altitudes—except when associated with the floor of Class B airspace. If navigational landmarks are not visible at night the procedure will be annotated: PROCEDURE NOT AUTHORIZED AT NIGHT.

CVFPs are not instrument approaches and do not have missed approach segments. They normally begin within 20 miles of the airport when published weather minimums prevail. ATC will authorize the procedure when the pilot reports a charted landmark or preceding aircraft insight. If the pilot is instructed to follow the preceding aircraft, the pilot is responsible for maintaining safe separation and wake turbulence interval. If at any point the pilot is unable to continue the approach or looses sight of the preceding aircraft, ATC will coordinate a go-around.

Figure 6-7 contains the San Jose International, San Jose, Calif., FAIRGROUNDS VISUAL RWY 30L and the Monterey Peninsula, Monterey, Calif., MOSS LANDING VISUAL RWY 28L charted visual approach procedures. Like DPs and STARs, CVFPs contain a plan view, along with a route description; appropriate communications frequencies are provided. Although each chart has a unique procedure name, CVFPs, unlike DPs and STARs, do not contain a computer code and are not filed as part of the flight plan.

Refer to the FAIRGROUND VISUAL RWY 30L procedure in Fig. 6-7. The first thing a pilot should note is that weather minimums for this procedure are a 2500-ft ceiling and 5 miles visibility. (Current weather and approach in use are advertised on the ATIS.)

Notes indicate that radar and the use of the SJC VOR/DME are required. Should either the radar or the VOR/DME be out of service, this procedure would not be assigned. The procedure description informs turbojet pilots approaching from the west-northwest to expect radar vectors in the vicinity of the cement plant for the visual approach. Aircraft should turn base no closer than the SJC R-170; plan view recommend altitude: "At or Above 5000, and final no closer than 6.6 DME at or above 2000 ft for noise abatement." An additional note advises pilots that lateral and vertical guidance on the final approach segment is available using the I-SJC localizer and glide slope (GS angle 3.00°).

Refer to the MOSS LANDING VISUAL RWY 28L procedure in Fig. 6-7. This procedure is not authorized at night or when the control tower is closed. Radar is required and the weather minimums are 3500-ft ceiling and 5 miles visibility. Vertical guidance on final is provided through a VASI with a glide angle of 3.50°. Pilots are referred to the enroute low-altitude chart L-2 and San Francisco sectional for additional guidance and information. Finally, pilots are advised that, because of practice approaches in VFR conditions, the runway 28L localizer may not always be activated—even though runways 28 are in use. Why would this be the case? The Monterey ILS RWY 10R localizer frequency is 110.7 MHz. The LOC DME RWY 28L frequency is also 110.7. Both facilities operating at the same time would result in frequency interference.

When you are assigned a CVFP, it's often helpful to have the appropriate instrument approach chart available with final approach guidance tuned in and identified. Especially in the busy terminal environment, it's easy to become distracted, or even momentarily disoriented. It's not unheard of for pilots to land on the wrong runway, or even at the wrong airport when flying visual procedures.

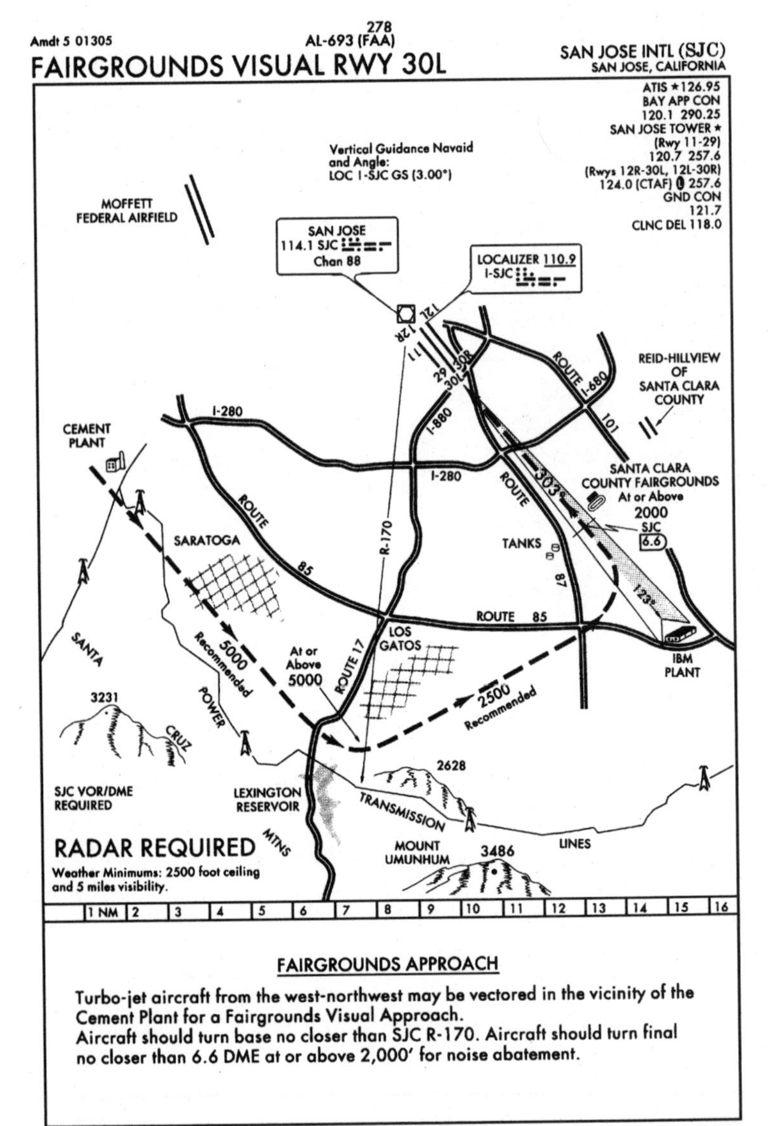

6-7 *Charted visual flight procedures have been developed to address environmental and noise concerns, provide a safe descent and approach course, and expedite the flow of air traffic.*

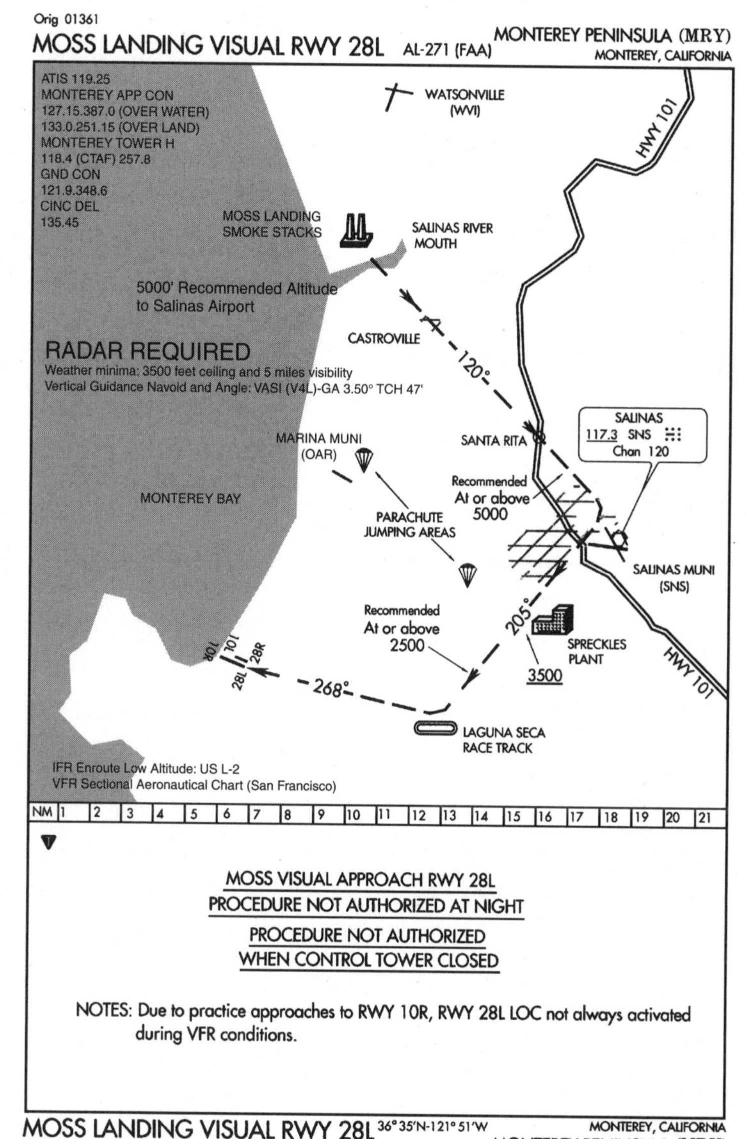

6-7 (*Continued*).

When destination weather is forecast to be at or above CVFP minimums, pilots can expect to be issued the procedure. However, it is the pilot's decision to accept a CVFP, or request an appropriate instrument approach procedure. If for any reason a pilot is not comfortable with a visual procedure—a CVFP or visual approach—upon request, ATC will assign an instrument approach procedure.

Some final thoughts on computer-filed flight plans. Pilots filing through DUAT or other computer systems will have their flight plans computer-checked for completeness and accuracy. Correct aircraft types, LOCIDs, and complete information will be required.

Case Study

I had a DC-8 pilot file for MKC (Kansas City Downtown Airport). I asked if he meant the Kansas City International Airport (MCI). He did. If this pilot had filed through DUAT, it would have resulted in considerable confusion.

Pilots can file, amend, or cancel a flight plan up to 1 hour before proposed departure time, but only through a compatible DUAT computer terminal. At this point ATC does not have access to the flight plan. One hour before proposed departure time, the flight plan is transmitted to the departure airport's ARTCC. Flight plan information transmitted to the center will contain only information necessary for the issuance of an air traffic clearance; any amendments will now have to be made through an FSS. Incorrect aircraft call signs or location identifiers will result in lost flight plans.

Case Study

The pilot filed a departure point of BFI (Boeing Field, Seattle) instead of BFL (Bakersfield, California)—both perfectly good location identifiers—the flight plan was sent to Seattle Center. When the pilot requested clearance, Bakersfield Tower had no information on the flight plan; nor any way of obtaining the information.

Close only counts with hand grenades and atomic bombs.

At times, even some of those in the FAA can go a little overboard.

Case Study

At an accident prevention meeting some years ago, an FAA FSDO representative insisted that pilots file separate flight plans from the destination to alternate in case of radio failure and a missed approach; please do not do this. It serves no practical purpose and only congests computer systems that are already nearly saturated.

7

Instrument approach procedure (IAP) charts

With the end of the Soviet Union and the Gulf War, the Global Positioning System (GPS) became available for civilian use. Testing of GPS approaches began in 1993, with the first stand-alone GPS approaches published in 1994. With the rapid adaptation of GPS the United States Coast Guard, which runs the loran-C radio navigation system, proposed in 1994 to turn off the aid after 1996, and again in 2001. When numerous aviation groups protested, the decommissioning of loran-C has thus far been postponed. At the present it appears loran will be with us at least for the near future, but without the nonprecision approaches originally envisioned.

Data required to execute a descent from the IFR enroute environment to a point where a safe landing can be made are presented on instrument approach procedure (IAP) charts. NACO IAP charts are contained in the terminal procedures publication (TPP), as discussed in Chap. 6. Changes that occur between revision cycles are advertised as FDC NOTAMs and incorporated in the NOTAM publication described in Chap. 1. The National Imagery and Mapping Agency (NIMA), Canada, and private vendors also produce instrument approach procedure charts. NIMA provides instrument approach procedure charts—which can be purchased through the FAA—for those areas they support with enroute charts.

IAPs are established by the FAA after careful analysis of obstructions, terrain, and navigational facilities. Procedures authorized by the FAA are published in the *Federal Register* as rule-making action under 14 CFR Part 97, which covers standard instrument approach procedures. Using this information, NACO and other charting agencies publish IAPs. 14 CFR 91.175 (a), regarding instrument approaches to civil airports states: "Unless otherwise authorized by the administrator, when an instrument letdown to a civil airport is necessary, each person operating an aircraft...shall use a standard instrument approach procedure prescribed for that airport in Part 97 of this chapter." Appropriate maneuvers, which include altitude, courses, and other limitations, are prescribed in these procedures. Many years of accumulated experience show that IAPs provide a safe letdown during instrument flight conditions. It is important that all pilots thoroughly understand these procedures and their use.

Pilots who use NACO chart material must be familiar with the TPP. This chapter discusses the following sections and charts included in the terminal Alaska and terminal procedures publications:

- Inoperative components table
- Explanation of terms/landing minima format
- Index of terminal charts and minimums
- IFR takeoff and departure procedures (except DPs)
- IFR alternate minimums
- General information and abbreviations (a list of abbreviations is contained in App. A)
- Chart legends
- Frequency pairing
- Radar minimums
- Terminal charts
- Rate of descent table

The TPP contains an index of terminal charts and minimums. Procedures are listed alphabetically by city name and airport name within the publication. If an airport has nonstandard takeoff minimums or a published departure procedure, the pilot is referred to Sec. C, "IFR Takeoff Minimums and Departure Procedures," of the TPP. Should the airport have nonstandard alternate minimums, the pilot is referred to Sec. E, "IFR Alternate Minimums," of the TPP. STARs, contained in

Sec. P, "Instrument Approach Procedures," airport diagram, and DPs are listed by page number.

Section M of the TPP contains the frequency pairing and MLS channeling table. On most civil navigation receivers, when the VHF frequency is displayed, the equipment automatically tunes the paired UHF DME frequency; this table also converts MLS or TACAN channels to the associated VHF frequency.

Volumes are $5^{3}/4 \times 8^{1}/4$ in, in both bound and loose-leaf format, shrink wrapped, with four holes punched along the top. Binding mechanisms are available for sale from NACO or private vendors.

IAP charts depict all related navigational data, communications information, and airport layout. The scale on IAPs varies; usually it is 1:500,000 or 1:750,000. Individual charts are to scale, except where concentric rings appear. Features include:

- Related navigational data
- Communications frequencies
- Reporting points/fixes
- Transitional data
- Obstacles
- Major hydrography
- Terrain on impacted IAPs
- Minimum safe altitude
- Holding patterns
- Missed approach procedures
- Approach minima data
- Airport sketches
- Airport lighting information

Approach procedure charts

Refer to Fig. 7-1. As can be seen, NACO is in the process of changing the format of approach procedure charts. These changes are based on user and organizational recommendations to make the charts more user friendly. The IAP chart is divided into five general sections:

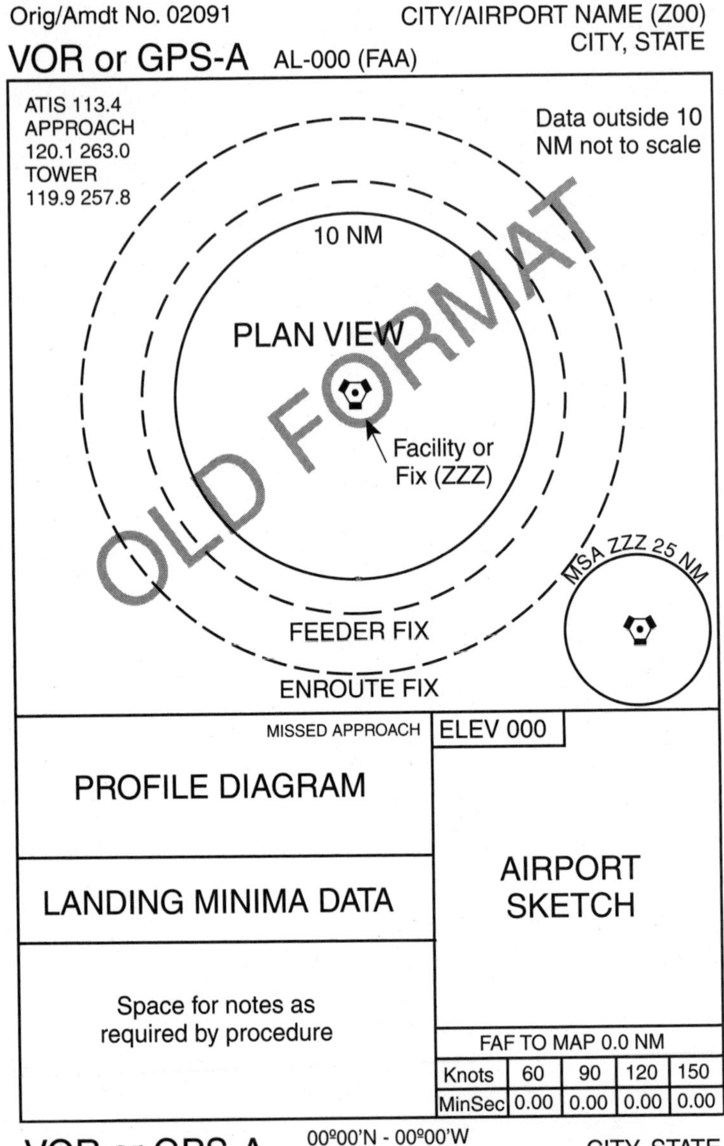

7-1 *Instrument approach procedure charts are divided into five sections: margin identification, plan view, profile diagram, landing minima data, and airport sketch.*

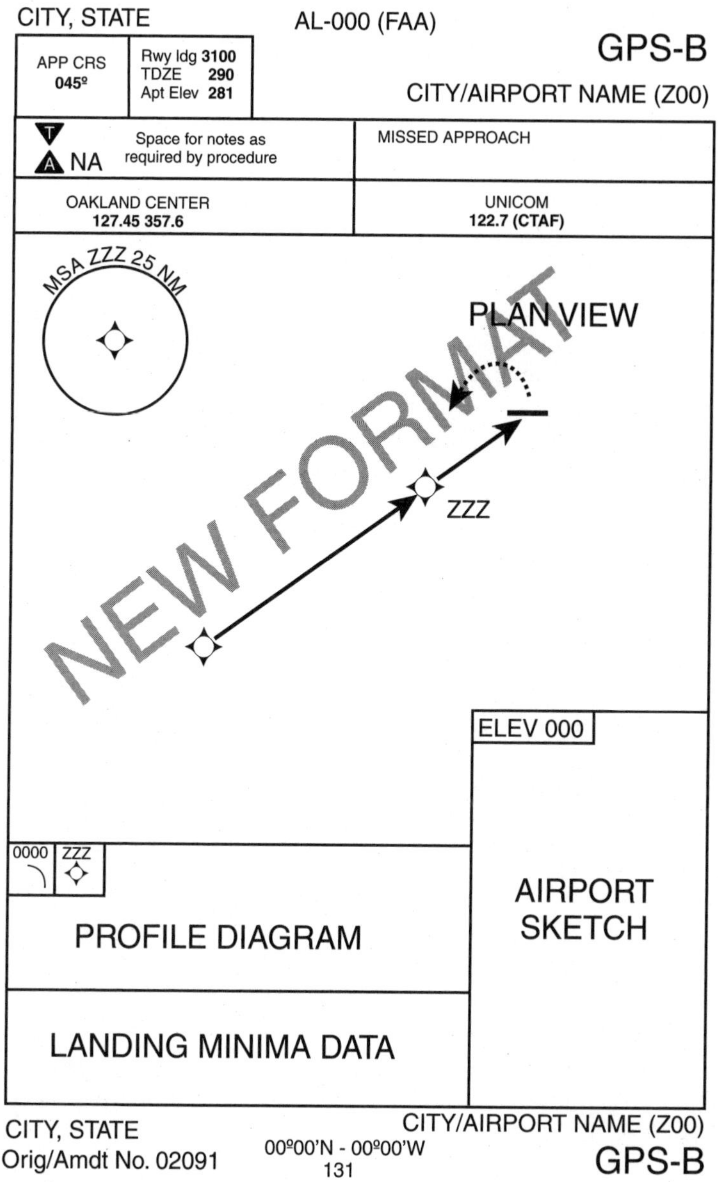

CITY, STATE
AL-000 (FAA)
GPS-B
APP CRS 045º
Rwy ldg 3100
TDZE 290
Apt Elev 281
CITY/AIRPORT NAME (Z00)
T
A NA
Space for notes as required by procedure
MISSED APPROACH
OAKLAND CENTER 127.45 357.6
UNICOM 122.7 (CTAF)
MSA ZZZ 25 NM
PLAN VIEW
NEW FORMAT
ZZZ
ELEV 000
0000
ZZZ
PROFILE DIAGRAM
AIRPORT SKETCH
LANDING MINIMA DATA
CITY, STATE
CITY/AIRPORT NAME (Z00)
Orig/Amdt No. 02091
00º00'N - 00º00'W
131
GPS-B

7-1 *(Continued).*

- Margin identification
- Plan view
- Profile diagram
- Landing minima data
- Airport sketch

The margin identification provides procedure identification, type, and number, along with other information. The plan view contains a bird's-eye look at the procedure. The profile diagram shows a side view of the approach. The landing minima data contains approach minimums information and procedure notes. The airport sketch provides an airport diagram, along with runway and airport information.

Bearings and courses are expressed in degrees magnetic. Radials are identified by the letter R (R-087) and expressed as magnetic bearing from the facility. Heights are expressed in feet above mean sea level (MSL) below 18,000 ft—except as noted, and as flight levels at and above 18,000 ft (FL180). Distances are nautical miles and tenths (10.4, ten point four nautical miles), except visibilities, which are expressed in statute miles and fractions ($2^1/_2$, two and one-half statute miles). Runway visual range (RVR) is expressed in feet (2400, two thousand four hundred feet). Aircraft speeds are in knots.

There are two basic types of instrument approach procedures: precision and nonprecision. A precision approach has some sort of electronic glide slope [ILS glide slope, precision approach radar, RNAV, GPS with wide area augmentation system (WAAS)]. A nonprecision approach does not provide a precision glide slope or glide path [localizer type directional aid (LDA), simplified directional facility (SDF), LOC, VOR, NDB, GPS, RNAV—without vertical guidance].

Margin identification

IAP identification is designed to be meaningful for the pilot, and permit ready identification to air traffic controllers. The procedure identification is derived from the type facility providing final approach course guidance (ILS, VOR, NDB, GPS, and the like). A solidus (/) indicates that more than one type of navigational equipment is required to execute the procedure (VOR/DME: The aircraft must be equipped with operational VOR and DME receivers.). When DME arcs and fixes are used in the procedure, but the approach identification does not include DME, the procedure may be used by aircraft not equipped with DME. In such cases arc transitions could not be used, and minimums might be higher. Where high altitude—jet penetration—procedures are available, the procedure identification is prefixed with the letters HI (HI-VOR RWY 5).

Note

With the implementation of IFR-approved RNAV and GPS systems, these navigation methods have been approved as substitutes for many ground navigational aids. Most are approved substitutes for VOR, ADF, and DME. This approval is contained in the aircraft's approved pilot's operating handbook (POH).

The microwave landing system (MLS) provides precision navigational guidance—azimuth, elevation, and distance—for exact alignment and descent of aircraft on approach to a runway. The MLS initially supplements and was supposed to eventually replace ILS as the standard landing system in the United States; however, the system requires special radio equipment aboard the aircraft that most owners have not installed because ILS facilities are more prevalent, among other reasons. With the planned implementation of precision GPS approaches it appears that MLS is a dead issue.

When any approach procedure meets criteria for straight-in landing (approach course within 30° of runway centerline), the runway number identifies the procedure. For example, "ILS RWY 14" indicates a straight-in approach landing on runway 14. This does not mean, however, that a circling approach may also be authorized using this basic procedure. A circling approach will be authorized only when the approach course is more than 30° from runway centerline. Additionally, even when the approach course is less than 30° from the runway, if a normal descent cannot be made for all categories, the approach will be labeled as circling. Circling approaches are identified by the type of navigational aid and an alphabetical suffix (VOR-A, NDB-C, GPS-B, and the like).

In cases where different approaches use the same final approach guidance to the same runway, a number differentiates the procedures (LDA/DME-1 RWY 18 and LDA/DME-2 RWY 18). In the preceding example, there are two LDA/DME approaches to runway 18 at the South Lake Tahoe Airport in California. The LDA/DME-1 has lower minimums than the LDA/DME-2 because the missed approach utilizes the Squaw Valley (SWR) VORTAC. Should SWR be inoperative, the procedure is not authorized; however, the LDA/DME-2, with higher minimums not utilizing SWR, could be executed with the SWR VORTAC out of service.

Effective January 25, 2001, the title of all RNAV instrument approach procedures changed. For example, RNAV (GPS) RWY 18 is an approved RNAV or GPS approach to runway 18. Controllers giving clearance for the RNAV (GPS) approach will not include the term GPS: "Cleared R-NAV runway 18 approach." At some locations more than one RNAV procedure may exist for the same runway. In such cases a letter preceding the runway identification will be used. For example, RNAV (GPS) Z RWY 18. (A few of the original VOR DME RNAV approaches remain. In such cases the approach will be labeled such. For example, the Ukiah, Calif., VOR DME RNAV or GPS-B approach. This approach is authorized for VOR DME RNAV equipment. Unless labeled such, other RNAV approaches are not authorized with this equipment.)

At airports where a program has been specifically approved, ATC might conduct simultaneous converging instrument approaches to runways having an included angle from 15° to 100°. These approaches are labeled CONVERGING. The procedures require separate instrument approach procedures for each converging runway. Missed approach points (MAP) must be at least 3 miles apart, and missed approach procedures must ensure that missed-approach-protected airspace does not overlap. Other requirements include radar, nonintersecting final approach courses, precision approach systems, and if the runways intersect, the controller must be able to apply visual separation and intersecting runway separation. Intersecting runways also require minimums of at least 700 ft and 2 miles. Straight-in approaches and landings are required. Whenever simultaneous converging approaches are in progress, pilots will be informed by the controller or automatic terminal information service (ATIS).

IAPs are labeled as original (Orig) or with an amendment number (Amdt 7), which helps determine chart currency. As shown in Fig. 7-1, on the old format charts this information is contained in the margin above and below the procedure title; on the new format charts this information is contained in the lower margin. FDC NOTAMs reference the original or amended chart: FDC...ILS RWY 7 AMDT 9. STRAIGHT-IN MINIMUMS NA; this FDC NOTAM changes the procedures for the ILS RWY 7 amendment 9 approach, and straight-in minimums for the approach are not authorized. If the approach chart is not ILS RWY 7, AMDT 9, it is obsolete. If this is a permanent change, the ILS RWY 7 AMDT 10 should contain the amended procedure.

In addition to procedure identification and chart amendment number, margins contain Julian date [the Julian date is 02091-this procedure was published (Orig) or revised (Amdt) on April 1, 2002], TPP page number, reference number and approving authority, airport name and location with three-letter identifier, and airport coordinates, as shown in Fig. 7-1.

As can be seen in Fig. 7-1, new format charts provide a "briefing strip" below the top margin identification. The briefing strip contains approach course bearing, runway information, airport notes, missed approach, and communications frequencies. For example, the new format chart in the example shows an approach course (045°), runway landing distance (3100), touchdown zone elevation (TDZE), and airport elevation (281, for quick reference).

Plan view

The plan view provides a horizontal picture of the procedure, plus communication frequencies, minimum safe altitudes within 25 nm of a central facility or fix, and denotes transition routes from enroute and feeder facilities. Note in

Fig. 7-1 that the new format chart places communication frequencies across the top of the page, in the briefing strip, just above the plan view.

At some airports a ground communication outlet (GCO) is provided. The communications data will contain the notation: "(when tower closed) GCO 123.975." Pilots at uncontrolled airports may contact ATC and FSS via a radio/telephone link. Pilots will use "four key clicks" to contact the ATC facility or "six key clicks" to contact the FSS.

Figure 7-2 contains plan view symbols unique to instrument approach procedure charts. Most are similar to those used on other instrument charts. Refer to the route data portion of Fig. 7-2. At the top of the box is a procedure track. The minimum altitude is 3100 ft. A procedure turn is not authorized (NoPT). The distance from the previous fix to glide slope (GS) intercept is 5.6 nm. The procedure bearing is 45° and the distance from the previous fix to the outer compass location (LOM) is 14.2 nm. At the bottom of the box is a transition or feeder route. The minimum altitude is 2000 ft, its bearing is 155, and the distance between fixes is 15.1 nm. Under fixes in Fig. 7-2 is an intersection, a DME arc with altitude (1700), the NAVAID that makes up the arc (ZZZ), and the distance of the arc (11 nm). Below are radials that define a fix, and lead-in radials (LR-198) or bearings (LB-198) to assist the pilot's transition from an arc to an approach course. At the bottom of the box, is an intersection (indicated by the short vertical line). The intersection name is ALPHA; it's based on the ZZZ NAVAID at 5 nm DME. Three course reversals are depicted: procedure turn, holding pattern, and teardrop.

An IAP might have up to four separate segments depicted in the plan view:

- Initial
- Intermediate
- Final
- Missed approach

The plan view is shown to scale within the 10-nm circle. Dashed concentric circles are used when information will

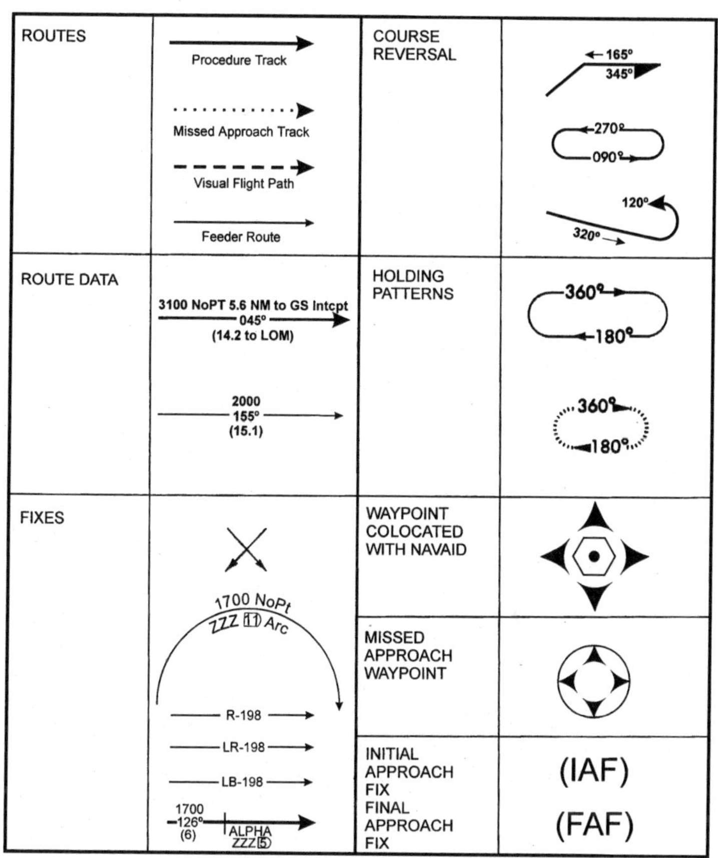

7-2 *Symbols on instrument approach procedure charts are similar to those used on other instrument charts.*

not fit to scale within the limits of the plan view area, labeled enroute facilities (fix) and feeder facilities (fix). Enroute facilities are NAVAIDs, fixes, and intersections that are part of the enroute low altitude structure; feeder facilities are NAVAIDs, fixes, and intersections used by ATC to direct aircraft to intermediate facilities, or fixes between the enroute structure and the initial approach segment.

When it is necessary to reverse direction to establish the aircraft inbound on an intermediate or final approach course, a course reversal or procedure turn is prescribed, symbolized in Fig. 7-2. A procedure turn is required, except when one of four conditions applies:

- The symbol NoPT is shown.
- Radar vectoring is being provided.
- A 1-minute holding pattern or a tear drop is published in lieu of a procedure turn.
- The procedure turn is not authorized.

The altitude prescribed for the procedure turn is the minimum altitude until the aircraft is established on the inbound course. The maneuver must be completed within the distance specified in the profile diagram. A barb indicates the direction or side of the outbound course on which the course reversal is to be made. Headings are provided for the 45° procedure turn; however, the point at which the turn may be commenced, and type and rate of turn, are left to the discretion of the pilot.

For example, the 45° procedure turn, racetrack pattern, teardrop procedure, or 80°–260° course reversal are acceptable. When a teardrop procedure or holding pattern replaces the procedure turn, the pilot must accomplish the published procedure, unless otherwise authorized by ATC. The absence of the procedure turn barb in the plan view indicates the procedure turn is not authorized.

When an approach course is published on an ILS procedure that does not require a procedure turn (NoPT), the following applies. In the case of a dogleg track where no fix is depicted at the point of interception on the localizer course, the total distance is shown from the facility or fix to the LOM. The minimum altitude applies until glide slope intercept, at which point the aircraft begins descent. When the glide slope is not utilized, the minimum altitude is maintained to the LOM.

Minimum safe altitudes (MSAs) provide at least 1000 ft obstacle clearance for emergency use within 25 miles of the procedure navigation facility. These will normally be the VOR or NDB, or final approach fix (FAF) for ILS and LOC approaches. Altitudes are rounded to the next higher 100-ft increments. These are illustrated in Fig. 7-1. As necessary, the 25-nm circle is divided into sections, with indicated MSA altitudes.

Emergency safe altitudes, normally used only in military procedures, are established with a 100-mile radius and feature a common altitude for the entire area, providing 2000 ft of obstacle clearance in designated mountainous areas.

Terrain will be added to the plan view of IAPs on a phased-in schedule over the next several years according to the following criteria:

- If the terrain within the plan view exceeds 4000 ft above airport elevation.
- If the terrain within a 6-nm radius of the airport reference point (ARP) rises to at least 2000 ft above airport elevation.

Approximately 235 airports in the United States will be affected. The plan view will be similar to existing charts using contour values and spot elevations, except that shaded contours will be depicted in varying shades of brown.

Profile diagram

The profile diagram contains minimum altitudes, provides maximum distance for procedure turns, altitudes over fixes, distances between fixes, glide slope angle for precision approaches, and missed approach procedures on old format charts. As illustrated in Fig. 7-1, new format charts have the missed approach narrative at the top of the chart, in the briefing strip. The profile diagram contains a graphic depiction of the missed approach procedure.

Figure 7-3 contains profile diagram symbols. When a navigational facility (VOR, NDB, marker beacon, and the like) establishes the fix, a solid vertical line is depicted. An intersection fix is shown as a dashed vertical line. Mandatory, minimum, maximum, and recommended altitudes are depicted in the same manner to other instrument charts.

Note the Maltese cross and lightning bolt symbols used to identify the final approach fix (FAF) or final approach point (FAP). (There is a subtle difference between the definition of the FAF and FAP. However, it has no operational significance to the pilot.) The precision approach glide slope intercept altitude is the minimum altitude for glide slope interception

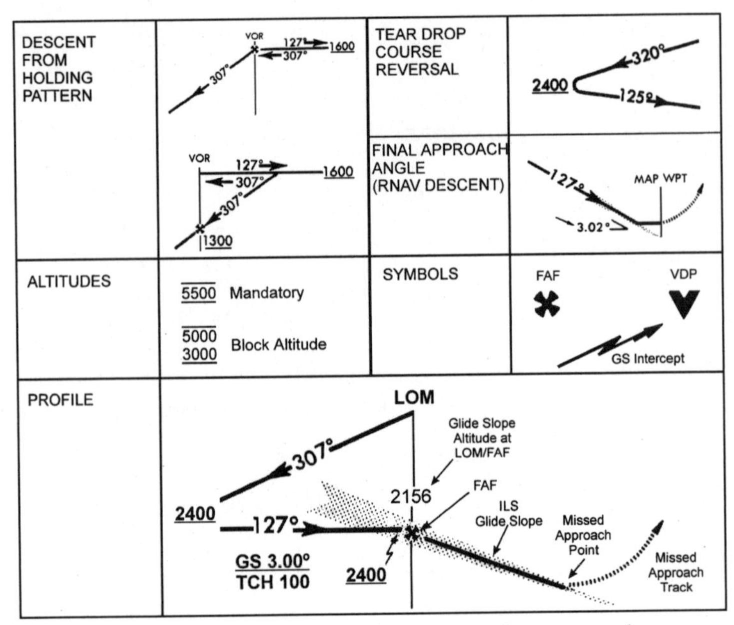

7-3 *The profile view provides the pilot with a vertical, or top to bottom, depiction of the approach.*

after completion of the procedure turn. The intercept altitude applies to precision approaches and, except where otherwise prescribed, becomes the minimum altitude for crossing the final approach fix if the glide slope is inoperative or not used. Glide slope angle (GS 3.00°) and threshold crossing height (TCH 100) are also depicted.

Stepdown fixes in nonprecision procedures might be provided between the final approach fix and the airport for the purpose of allowing a lower minimum descent altitude after passing an obstruction. Normally, there is only one stepdown fix between the final approach fix and the MAP. If the stepdown fix cannot be identified for any reason, the altitude at the stepdown fix becomes the minimum descent altitude (MDA). Final approach angles for aircraft equipped with RNAV allow the pilot to determine the appropriate descent rate (3.02°). The visual descent point (VDP) on a nonprecision straight-in approach is where normal descent from the minimum descent altitude to the runway touchdown point begins.

The inside back cover of the TPP contains a rate of descent table that provides descent rates in feet per minute for planning and executing precision descents. Table 7-1 contains an excerpt from the TPP's rate of descent table. Descent rate is based on glide slope angle and ground speed. (An advantage of RNAV systems is a constant, direct readout of ground speed.) The left column contains vertical path angle in degrees and tenths. The next column shows descent gradient in feet per nautical mile. The top row shows aircraft ground speed in knots.

For example, let's say the profile diagram shows an ILS glide path angle of 3.00°. The pilot plans a 100-knot IAS approach, with reported surface winds straight down the runway at 10 knots. If DME or GPS were available, ground speed would be known; however, under the given conditions we would estimate ground speed at 90 knots (100 − 10 = 90). Considering a glide angle of 3°, at a ground speed of 90 knots, from Table 7-1 we see the approximate rate of descent required would be 480 feet per minute. (ATC normally provides the pilot with approximate rate of descent information on precision radar approaches.)

Landing minimum data

Figure 7-4 shows landing minima data. Aircraft approach categories are based on approach speed. Landing minima are established for six aircraft approach categories: A, B, C, D, E, and COPTER. (In the absence of COPTER MINIMA, helicopters may use category A minimums.) Where the airport landing surface is not adequate, or other restrictions prohibit

Table 7-1 Rate of descent table

Angle of descent	Feet/nm	90	105	120	135
2.8	297	446	520	594	669
2.9	308	462	539	616	693
3.0	318	478	557	637	716
3.1	329	494	576	658	740
3.2	340	510	594	679	764

certain categories of aircraft, a NOT AUTHORIZED (NA) notation appears. Approach category E is normally published only on military high-altitude procedures for aircraft with approach speeds in excess of 165 knots.

The same minimums apply to day and night operations, unless different minimums are specified at the bottom of the box in the space provided for symbols or notes. Minimums for full ILS and LOC only, straight in, and circling appear directly under each aircraft category. When there is no division line between minimums for each category, the minimums apply to two or more categories. Normally, when a stepdown fix is within 3 miles of the airport, dual equipment will be required to establish the fix. Under these conditions two sets of minimums appear, one will be labeled DUAL VOR or DME MINIMUMS.

Circling approach obstacle clearance areas are based on aircraft approach category. Pilots must keep in mind that an aircraft that falls in category A, but is circling to land at a speed

CATEGORY	A	B	C	D
SPEED KT	0-90	91-120	121-140	141-165

CATEGORY	A	B	C	D
S-ILS 27	1352/24 200 (200-½)			
S-LOC 27	1440/24 288 (300-½)			1440/50 288 (300-1)
CIRCLING	1540-1 361 (400-1)	1640-1 461 (500-1)	1640-1½ 461 (500-1½)	1740-2 561 (600-2)

CATEGORY	COPTER
H-176°	1640-1 363 (300-½)

7-4 Landing minimum data is subdivided by aircraft approach category, approach aid utilized (ILS or LOC), and a straight-in or circling maneuver.

in excess of 91 knots, must use the approach category B minimums when circling to land. Minimums are based on obstruction clearance, as shown in Table 7-2, because higher speeds require consideration of larger obstruction clearance areas. That's why MDAs and visibilities typically increase with higher approach speeds. This explains the following FDC NOTAM:

FDC 1/1521 SCK FI/T /SCK/ STOCKTON METROPOLITAN, STOCKTON, CA. ILS RWY 29R AMDT 18..VOR RWY 29R AMDT 17..NDB RWY 29R AMDT 14: CATS B/C CIRCLING MDA 580/HAA 550. 265 FT MSL CRANE 1.4 NM WEST OF AER 11L.

The circling MDAs for aircraft categories B and C have been raised because of a 265-ft MSL crane 1.4 miles west of approach end of runway (AER) 11L. Category A is not affected because the crane is outside of the obstruction clearance radius. Category D minimums are already at 580 ft because of other obstructions; therefore, this FDC NOTAM applies only to category B and C operations. To decode FDC and other NOTAMs refer to App. A.

Landing minimums are expressed as minimum descent altitudes (MDA) or decision height (DH), and visibility in statute miles or runway visual range (RVR) in feet. MDA means the lowest altitude, expressed in feet above mean sea level, to which descent is authorized on final approach, where no electronic glide slope is provided, or during circle-to-land maneuvering. DH means the height at which a decision must be made during an approach with an ILS/MLS or precision approach radar (PAR) or RNAV vertical navigation (VNAV) to (1) continue the approach only if the runway environment is

Table 7-2 Circling approach obstacle areas

Approach category	Radius, nm
A	1.3
B	1.5
C	1.7
D	2.3
E	4.5

visible or (2) immediately execute a missed approach. This height is expressed in feet above mean sea level and as a radar altimeter setting (RA) for Category II and III procedures, which require additional aircraft equipment and aircrew certification. Height above airport (HAA) and height above touchdown (HAT) are also published. HAA indicates the height of the MDA above the published airport elevation, published in conjunction with circling minimums for all types of approaches; HAT indicates the height of the DH or MDA above the highest runway elevation in the touchdown zone, for the first 3000 ft of runway, and is published in conjunction with straight-in minimums.

Refer to Fig. 7-4. In this example airport elevation is 1179 ft, and runway touchdown zone elevation is 1152 ft. For a straight-in full ILS approach to runway 27 (S-ILS 27), minimums for categories A through D are the same. Since this is a precision approach, the DH is 1352 ft MSL. Required landing visibility is an RVR of 2400 ft (24). The next number, 200, is the HAT. Values contained within parentheses apply only to military operations.

Should the glide slope be inoperative or the aircraft not equipped with a glide slope receiver, straight-in localizer minimums would apply (S-LOC 27). Since this is a nonprecision approach, the MDA for categories A through D is 1440 ft. Notice, however, that the visibility requirement for category D has been increased to an RVR of 5000 ft. These are followed by CIRCLING minimums. Minimums increase progressively for higher-speed approach categories. Since RVR only applies to a specific runway, landing visibilities are shown in whole and fractions of statute miles.

Note

Should the glide slope be inoperative or the pilot unable to execute a full ILS approach, localizer minimums apply. However, the pilot should request and ATC will provide clearance for "ILS RUNWAY 27" approach. For a circling approach the clearance again would be for "ILS RUNWAY 27 CIRCLE TO LAND RUNWAY 9." Appropriate minimums would apply.

Copter minimums are also shown in Fig. 7-4. The copter approach direction is a heading of 176° (H-176°). Otherwise, the MDA or DH and visibilities are the same, with 363 the height of the MDA or DH above landing area (HAL).

Table 7-3 shows a visibility conversion from RVR to statute miles. To authorize RVR minimums, the procedure must have, in addition to basic components, RVR reported for the runway, high-intensity runway lights (HIRL), and all-weather runway markings for precision procedures, or instrument runway markings for nonprecision approaches. If RVR minimums for takeoff or landing are published, but RVR equipment is inoperative, RVR must be converted and applied as ground visibility as shown in Table 7-3.

Circling minimums provide adequate obstruction clearance. The pilot must not descend below this altitude until in a position to make the final descent for landing. The pilot must determine the exact maneuver for this procedure, based on airport design, aircraft position, altitude, and airspeed. This requires good judgment and a knowledge of aircraft capabilities. In general, the following basic rules should be applied. Maneuver the shortest path to base or downwind considering weather conditions. There is no restriction from passing over the airport or other runways. Keep in mind, especially at uncontrolled airports, that many circling maneuvers may be made while VFR flying is in progress. It might be desirable to overfly uncontrolled airports to determine wind and turn indicators, and to

Table 7-3 RVR/visibility comparable values

RVR, ft	Visibility, sm
1600	1/4
2000	3/8
2400	1/2
3200	5/8
4000	3/4
4500	7/8
5000	1
6000	1 1/4

observe other traffic. Standard left turns or controller instructions must be considered.

The instrument pilot must know which category applies to the aircraft to be flown. Most often FSS controllers are not aware of aircraft approach categories. In such cases pilots can expect to receive information that may not apply; however, as always, it's the pilot's responsibility to determine whether the information will affect their operation.

ATC might authorize a side-step maneuver to an adjacent parallel runway when separated by 1200 ft or less. When so cleared, pilots are expected to commence the side-step maneuver as soon as possible after the runway or runway environment is sighted. Landing minimums to adjacent runways will be higher than those for the primary runway, but will normally be lower than circling minimums.

Category of operation, with respect to the operation of aircraft, means a straight-in ILS approach to a runway under special instrument approach procedures issued by the FAA or other authority. Category II and Category III operations permit lower landing minimums for specially trained crews operating specially equipped aircraft with radios that receive specially certificated ILS systems; the DH is expressed as a radar altimeter setting [RA 100 (feet)] as opposed to a mean sea level altitude.

Space is provided below the landing minima data box for notes as required by the procedure on old format charts. On the new format charts, as shown in Fig. 7-1, notes are provided at the top of the chart in the briefing strip. Notes contain additional information about the procedure, such as altimeter setting source, runway lighting activation procedures, and nonstandard alternate and takeoff minimums, or published departure procedure.

14 CFR Part 91.169 (c), regarding IFR alternate airport weather minimums, specifies standard alternate criteria. "Precision approach procedure: Ceiling 600 feet and visibility 2 statute miles. Nonprecision approach procedure: Ceiling 800 feet and visibility 2 statute miles." Alternate minimums require both ceiling and visibility. A black triangle enclosing a white letter

A indicates that alternate minimums are not standard. If the airport is not authorized for use as an alternate, the letters NA follow the symbol—this is illustrated on the new format chart in Fig. 7-1.

When a pilot elects to proceed to the selected alternate airport, the alternate ceiling and visibility minimums no longer apply; published landing minimums now apply to the new destination. The alternate airport becomes the new destination, and the pilot uses the landing minimum appropriate to aircraft equipment, approach speed, and the type of procedure selected.

A black triangle enclosing a white letter T indicates nonstandard takeoff minimums or published departure procedures, or both. 14 CFR Part 91.175 (f), on civil airport takeoff minimums, specifies standard takeoff minimums for pilots operating under FAR Part 121, 125, 127, 129, or 135, which apply to air carrier, air taxi, and large aircraft operators. Nonstandard alternate and takeoff minimums and departure procedures are tabulated in the TPP.

Landing minimums published in the landing minima data section of IAP charts are based on the full operation of all electronic components and visual aids, such as approach lighting systems, associated with the procedure. Higher minimums apply when visual aids or RVR is inoperative. The inoperative components table of the TPP is published on the inside front cover. Inoperative components result in increased visibility requirements. Why? The purpose of the approach lighting system is to allow the pilot to visually acquire the landing environment at a greater distance from the runway. Inoperative components are contained in Table 7-4.

For example, a pilot flying an aircraft in approach category B is planning an ILS approach, with visibility minimum of 1/2 mile, to an airport where the approach lighting system (MALSR) is inoperative. According to the table in the TPP, inoperative MALSR requires an increase of 1/4 mile visibility; therefore, the required visibility is now 3/4 of a mile.

Individual IAP charts might contain notes that increase minimums and supersede those contained in Table 7-4; however,

Table 7-4 Inoperative components table

Inoperative component or aid	Approach category	Increase visibility
ILS, MLS, and PAR		
ALSF 1 and 2, MALSR, and SSALR	ABCD	1/4 miles
ILS with visibility minimum of 1800 RVR		
ALSF 1 and 2, MALSR, and SSALR	ABCD	To 4000 RVR
TDZL and RCLS	ABCD	To 2400 RVR
RVR	ABCD	To 1/2 mile
VOR, VOR/DME, LOC, LOC/DME, LDA, LDA/DME, SDF, SDF/DME, GPS, RNAV, and ASR		
ALSF 1 and 2, MALSR, and SSALR	ABCD	1/2 mile
SSALS, MALS, and ODALS	ABC	1/4 mile
NDB		
ALSF 1 and 2, MALSR, and SSALR	C	1/2 mile
ALSF 1 and 2, MALSR, and SSALR	ABD	1/4 mile
MALS, SSALS, ODALS	ABC	1/4 mile

ILS glide slope inoperative minimums are always published as localizer minimums on IAP charts. The following general rules commonly apply when components are inoperative:

- Runway lights are required for night operations; if inoperative, the procedure is not authorized.
- When the facility providing course guidance is inoperative, the procedure cannot be used.
- If more than one component is specified in the procedure identification (VOR/DME), when either the VOR or DME is inoperative, the procedure is not authorized.
- Localizer minimums apply when the ILS glide slope is inoperative or not used.
- Compass locator or precision radar may be substituted for the ILS outer or middle marker.
- Surveillance radar may be substituted for the ILS outer marker.
- DME, at the glide slope site, may be substituted for the outer marker when published on the ILS procedure.

Case Study

In March of 2001 a chartered Gulfstream III crashed in Aspen, Colo. All on board were killed. The crash occurred in light snow about 33 minutes after official sunset while the crew was trying to intercept the instrument approach. Night instrument approaches to this airport are not authorized. The National Transportation Safety Board (NTSB) noted that the sun would have set below the mountainous terrain about 25 minutes before the official sunset time. The investigation prompted the following recommendation: The FAA should clarify the definition of "night" to take into account terrain and other variables.

14 CFR Part 1 defines night as "the time between the end of evening civil twilight and the beginning of morning civil twilight, as published in the *American Air Almanac*." The *American Air Almanac* was last published in 1956. Civil twilight ends when the sun is 6° below the horizon; it is still light enough to carry on ordinary outdoor activities without artificial

light. Six degrees below the horizon occurs at approximately 30 minutes after sunset and 30 minutes before sunrise. In the previous case study, the weather and terrain would have certainly reduced the amount of ambient light.

Facilities that establish a stepdown fix (fan marker, VOR radial, and the like) are not components of the basic approach procedure. Additional methods of identifying a fix may be used when published on the procedure.

Radar minimums for airports with radar instrument approaches appear in Sec. N of the TPP. Virtually all published radar minimums are for joint-use, civil and military, airports. Although most airports with surveillance radar have emergency, unpublished procedures. Table 7-5 shows the radar instrument approach minimums format used in the TPP. ATC provides course, altitude, and missed approach guidance; therefore, only landing minimum data, in a slightly different format from IAPs, is provided. DH is shown for precision approach radar (PAR) and MDA for airport surveillance radar (ASR), along with visibility. Minimums in parentheses are applicable only to military operations. When the pilot is cleared for an approach and instructed to circle to another runway, the circling MDA and weather minima apply.

Airport sketch/airport diagram

An airport sketch is designed to assist pilots transiting from instrument to visual conditions. The airport sketch is located in the lower right (sometimes left) corner of the IAP, as illustrated in Fig. 7-1. Airport diagrams are full-page illustrations of the airport complex. They are specifically designed to assist pilots while maneuvering on the ground by providing runway and taxiway configurations, and provide information for updating inertial navigational systems (INS).

Figure 7-5 shows airport sketch/diagram symbols. The scale perspective of airport diagrams and the northern orientation varies. The symbols provide runway surface information and runway and taxiway condition, and indicate the presence of displaced thresholds. Runway touchdown elevations are shown, and gradients too are shown when they exceed 0.3 percent.

Table 7-5 Radar instrument approach minimums

	RWY	CAT	DH/MDA-VIS	HAT/HAA	CEIL-VIS
PAR	8	ABCDE	4916-3/4	200	(200-3/4)
ASR	26	ABCDE	4900-1	274	(200-1)
CIRCLING		A	5100-1	384	(400-1)

Runway length is the length of the runway, end to end, including displaced thresholds, but excluding overruns. When a displaced threshold is depicted, a note is added to indicate runway available for landing. Availability of arresting gear and optical landing systems are noted for military operations.

The sketch also indicates airport elevation and time and distance from the FAF to the MAP; time and distance from FAF to MAP are omitted on ILS and GPS procedures when the NAVAID is collocated with the MAP or when DME is required for the procedure.

Pilots should be familiar with the lighting systems at destination airports, especially when operating close to minimums or at night. Airport lighting codes are contained in Fig. 7-6; specific approach lighting system (ALS) configurations and visual approach slope indicator (VASI) glide path configurations are contained in the *Aeronautical Information Manual* (AIM).

Old-format IAP charts display the ALS configuration and code adjacent to the instrument runway in the airport sketch. New-format IAP charts show the ALS configuration, code,

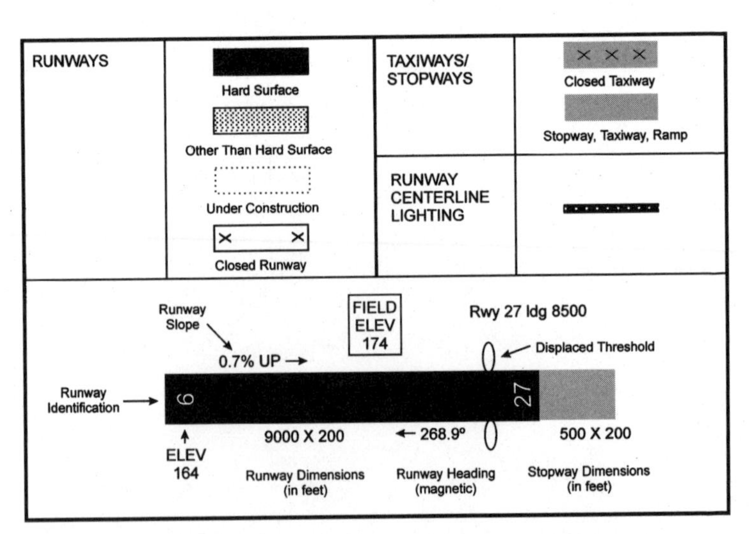

7-5 *Airport symbols describe the runway environment.*

APPROACH LIGHTING SYSTEM (ALSF-2)	A	2 BAR VISUAL APPROACH SLOPE INDICATOR (VASI)	V
APPROACH LIGHTING SYSTEM (ALSF-1)	A1	"T"-VISUAL APPROACH SLOPE INDICATOR ("T"-VASI)	V1
SHORT APPROACH LIGHTING SYSTEM (SALS/SALSF)	A2	PULSATING VISUAL APPROACH SLOPE INDICATOR (PVASI)	V2
SIMPLIFIED SHORT APPROACH LIGHTING SYSTEM (SSALR)	A3	3 BAR VISUAL APPROACH SLOPE INDICATOR (VASI)	V3
MEDIUM INTENSITY (MALS & MALSF) OR SIMPLIFIED SHORT ALS (SSAL & SSALF)	A4	TRI-COLOR VISUAL APPROACH SLOPE INDICATOR (TRCV)	V4
MEDIUM INTENSITY APPROACH LIGHTING SYSTEM (MALSR)	A5	ALIGNMENT OF ELEMENTS SYSTEM (APAP)	V5
OMNIDIRECTION-AL APPROACH LIGHTING SYSTEM (ODALS)		PRECISION APPROACH PATH INDICATOR (PAPI)	P

7-6 *Coded runway approach lighting system symbols indicate the availability and type of, and activation procedure for, runway lighting aids.*

and contractions (MALSR, etc.) in the briefing strip. A black dot at the top of the ALS code means that sequenced flashing lights are installed with the ALS. The absence of the dot, for example A3 in Fig. 7-6, indicates sequenced flashing lights are not available. A negative (white-on-black) symbol, for example A2 and V2 in Fig. 7-6, indicates pilot-controlled lighting (PCL). If PCL is nonstandard—indicated by a asterisk following the symbol, the pilot must refer to the *Airport/Facility Directory* for activation procedures.

VASI V is a 2-bar installation. The VASI may only be on one side of the runway, and may consist of 2, 4, or 12 lights. VASI

V3 is a 3-bar installation, with a threshold crossing height to accommodate long-bodied or jumbo aircraft. Like the 2-bar VASI, the 3-bar may be on only one side of the runway, and may consist of 6 or 16 lights. The alignment of elements systems (APAP) consists of painted panels, which may be lit at night. To use the system the pilot positions the aircraft so the elements are in alignment.

Detailed airport diagrams for large and complex airports are listed in Sec. B, "Index of Terminal Charts and Minimums," of the TPP. With detailed runway, taxiway, and facility names and locations, these diagrams are invaluable for pilots operating into or out of unfamiliar airports. Figure 7-7 contains an airport diagram for the Oakland, Calif., Oakland/Metropolitan Oakland International Airport.

In addition to the detail, including taxiway identification, the chart provides runway weight bearing capacity or pavement classification number (PCN)—for example, RWY 15-33: S125, T65, TT100. These can be decoded in the *Airport Facility/Directory,* covered in Chap. 8. (One additional symbol appears in Fig. 7-7. Note the two solid triangles at the end of runway 11. This indicates runway radar reflectors for military use.)

Using approach charts

Approach segments begin and end at designated fixes; however, under some circumstances certain segments might begin at specified points where no fixes are available. The fixes are named to coincide with the associated segment. For example, the intermediate segment begins at the intermediate fix and ends at the final approach fix, where the final approach segment begins.

An instrument approach begins at the initial approach fix (IAF). Feeder routes provide a transition from the enroute structure to the IAF; however, when the IAF is part of the enroute structure, there might be no need to designate feeder routes. This is the point where the aircraft departs the enroute phase and maneuvers to enter an intermediate segment. An initial approach may be made along an arc, radial, course,

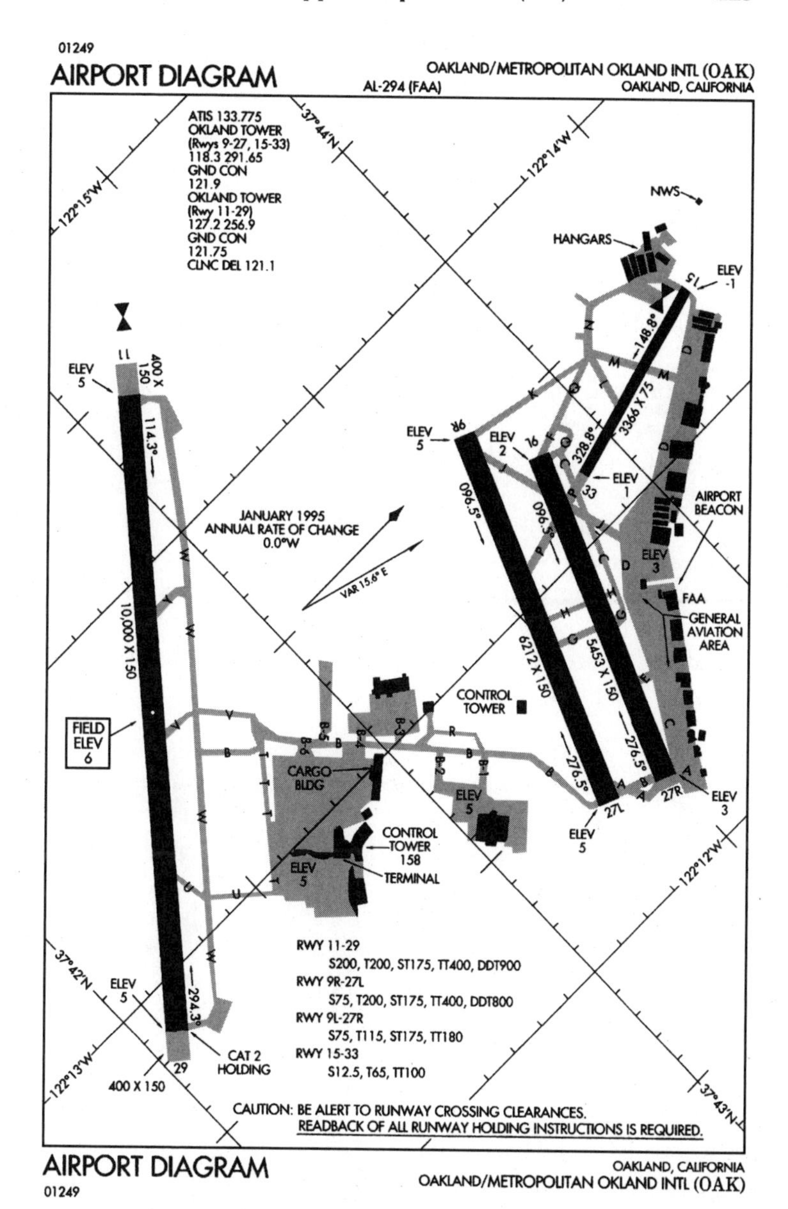

7-7 *Airport diagrams are issued for large, complex airports; radar approach minimums appear in a separate section of the terminal procedures publication.*

heading, radar vector, or any combination. Procedure turns, holding pattern descents, and high-altitude penetrations are initial segments. Positive course guidance is required, except when dead reckoning courses can be established over limited distances. Altitudes are established in 100-ft increments. Normally a minimum of 1000-ft obstacle clearance is provided.

The intermediate approach segment blends the initial approach into the final approach segment. In this segment, aircraft configuration, speed, and positioning adjustments are made for transition to the final approach. The intermediate segment begins at the intermediate fix (IF), or point, and ends at the final approach fix (FAF). There are two types of intermediate segments: the radial or course intermediate segment, and the arc intermediate segment. Altitudes are established in 100-ft increments. The optimum descent gradient is 250 feet per nautical mile, but where a higher descent gradient is necessary, 500 fpnm is allowed. Normally a minimum of 500-ft obstacle clearance is provided.

Alignment and descent for landing are accomplished in the final approach segment, which begins at the final approach fix. The FAF is designated on IAP charts with a Maltese cross; a lightning bolt symbol indicates glide slope or glide path intercept altitude and FAF for a precision approach. When a NAVAID, such as a VOR or NDB, is located on the airport, an FAF is not designated. The final approach point (FAP) on these procedures designates the point where the aircraft is established inbound on the final approach course from the procedure turn and the final approach descent commences. The FAP serves as the FAF and identifies the beginning of the final approach segment for a nonprecision approach. Descent gradients generally vary from 300 to 400 fpnm for nonprecision approaches. Obstacle clearance begins at the FAF or FAP and ends at the runway or missed approach point. Obstacle clearance for nonprecision approaches normally varies between 250 and 350 ft, depending upon the type of procedure (LOC, VOR, NDB, GPS, and the like). An aircraft should never descend below the published altitude on the final approach segment, except for final descent to landing. Glide slope on a precision approach is normally 3°, but can vary slightly on individual procedures; the glide path is normally 1° thick, which represents a vertical distance of approximately 920 ft when the aircraft is 10 miles from touchdown, narrowing to a few feet at touchdown. Obstacle clearance for precision approaches is based on a complicated formula; basically, the closer to the runway, the lower the obstacle clearance. The pilot should never allow a full-scale deflection below the glide slope while on the final approach segment.

Some nonprecision approaches have designated visual descent points (VDPs). When an approach incorporates a VDP, it is identified by a navigational fix. Where a visual approach slope indicator (VASI) is installed, the VDP is located at the point where the lowest VASI glide slope intersects the lowest MDA; where a VASI is not installed, the VDP is located at the point on the final approach course at the MDA, normally, where a descent gradient to the threshold of 300 to 400 fpnm begins.

Missed approach procedures are established for all instrument approaches. The missed approach procedure begins at the decision height for precision approaches, or at a specified point for nonprecision procedures. This segment is designed to be simple, specify an altitude, and whenever practical, provide a clearance limit. Whenever possible, the missed approach track will be a continuation of the final approach course. A turn of 15° or less is considered straight. Missed approach obstacle clearance begins from the missed approach point (MAP) and ultimately provides 1000-ft clearance. It is important for the pilot to immediately establish the aircraft in a positive rate of climb at the MAP. Obstacle clearance protected areas are predicated on the assumption that the missed approach is executed at the prescribed minimum descent altitude (MDA) or decision height (DH). If visual reference is lost at any point during a circling approach, the missed approach procedure must be immediately executed. To become established on the missed approach course, the pilot should make an initial climbing turn toward the landing runway and continue to turn until established on the missed approach course. Because the circling maneuver may be accomplished in more than one direction, different flight paths may be required to establish the aircraft on the missed approach course.

The MAP might be the intersection of an electronic glide path with a DH, a NAVAID, a fix, or a specified distance from the FAF. The specified distance will not be greater than the distance from the FAF to the runway, or prior to the VDP. The MAP for a full-ILS, microwave landing system (MLS), and precision approach radar (PAR) is at the DH; the localizer-only MAP is usually over the runway threshold. In some nonprecision procedures, the MAP might occur prior to reaching the runway threshold in order to clear obstructions in the missed approach climb. The pilot determines the MAP by timing from the FAF for some nonprecision approaches. The distance from FAF to MAP, and time and speed table, are provided below the airport sketch.

When the missed approach procedure specifies holding at a facility or fix, holding is accomplished in accordance with the depicted pattern on the plan view, and at the altitude in the missed approach instructions, unless otherwise specified by ATC. ATC might specify an alternate missed approach procedure.

Various terms are used in the missed approach procedure that have specific meanings with respect to climbs and turns, usually for obstruction avoidance:

- "Climb to…" means a normal climb along the prescribed course.
- "Climbing right turn…" means a climbing right turn as soon as safety permits, normally to avoid obstructions straight ahead.
- "Climb to 2,400 turn right…" means climb to 2400 ft prior to making the right turn, normally to clear obstructions.

A Category I ILS approach procedure provides an approach to a decision height of not less than 200 ft. The Category I ILS consists of the following components: localizer (LOC), glide slope (GS), outer marker (OM), and middle marker (MM). A complete Category I ILS consists of these components. When the localizer fails an ILS approach cannot be used. When the glide slope becomes inoperative, the ILS becomes a nonprecision approach. When other components become inoperative the ILS may continue to be used, sometimes with increased landing minimums. Nondirectional radio beacons, called compass locators, might be installed at the outer and middle marker sites, but are not considered as basic components of the ILS; however, when installed, they may be substituted for the outer or middle marker. DME might also be associated with an ILS. When installed with the ILS, DME may be substituted for the outer marker. When a unique operational requirement exists, DME information from a separate facility might also be used to provide DME arc initial approaches or a FAF for back-course approaches, or substituted for the outer marker. At one time, DME information from a separate facility was used to establish fixes inside the FAF. On several occasions fatal accidents resulted from pilots tuning the receiver to the wrong facility or misinterpreting the information. Pilots should use extra caution when obtaining DME information from a facility not being used for course guidance; positive aural identification of a navigational facility must be incorporated into every such frequency change.

An ILS is identified by the Morse code letter I preceding the airport location identifier; the ILS at the Stockton, Calif., airport (SCK), is identified ISCK. Compass locators normally have a two-letter Morse code identifier. The outer compass locater (LOM) uses the first two letters of the airport identifier (SC). The middle compass locator (LMM), when installed, uses the last two letters of the airport identifier (CK). Most LOMs are given five-letter names, like intersections. Because large airports have more than one ILS procedure, random three-letter identifiers, preceded by the letter I, are used. Microwave landing systems are identified by the letter M.

Approach control informs a pilot which approach to expect when the airport has more than one instrument procedure, perhaps an IAP or CVFP. This information is provided by the controller or broadcast on the ATIS. When the advertised approach cannot be executed, or another procedure is desired, it must be specifically requested by the pilot.

Case Study

Years ago I accompanied an instrument student in a Mooney on a flight to San Diego's Lindbergh Field. The approach in use was the localizer back-course, which at the time required the use of a marker beacon receiver. The Mooney was equipped only with VOR/localizer and ADF receivers. We planned to execute a front-course localizer approach, utilizing the ADF to establish outer and middle marker locations. ATC suggested and we accepted a radar vector below the overcast, over the ocean because of traffic in the area. Ground control subsequently instructed us to call approach control on the phone; I explained the situation and was relieved when the ATC supervisor replied, "Oh, OK."

Some say that every instrument approach procedure is different. I believe that this is like saying, "Your glass is half empty." In most respects every approach is just like every other. They simply consist of courses and altitudes to be maintained to a point where either a landing or missed approach is performed.

During the flight planning stage a pilot needs to review all destination procedures. This may be as simple as one IAP or as complex as more than a dozen. Let's assume that our destination is the Tracy Municipal Airport, Tracy, Calif. A review of the index in the TPP shows five procedures: two NDB, two GPS, and the VOR or GPS-A. Let's assume we do not have ADF or GPS equipment. Therefore, we will need to execute the VOR approach. The Tracy VOR procedure is illustrated in Fig. 7-8.

Our first task is to plan a route to an IAF for the approach. From the plan view we see that the Manteca (ECA) VOR is the only IAF for this approach. Therefore, we would plan our route to end at ECA. From ECA we would simply file our destination Tracy airport (TCY) "...ECA TCY." Be careful. As shown in the margin the identifier for the airport is TCY, not to be confused with the intersection of the same name, but identifier TRACY.

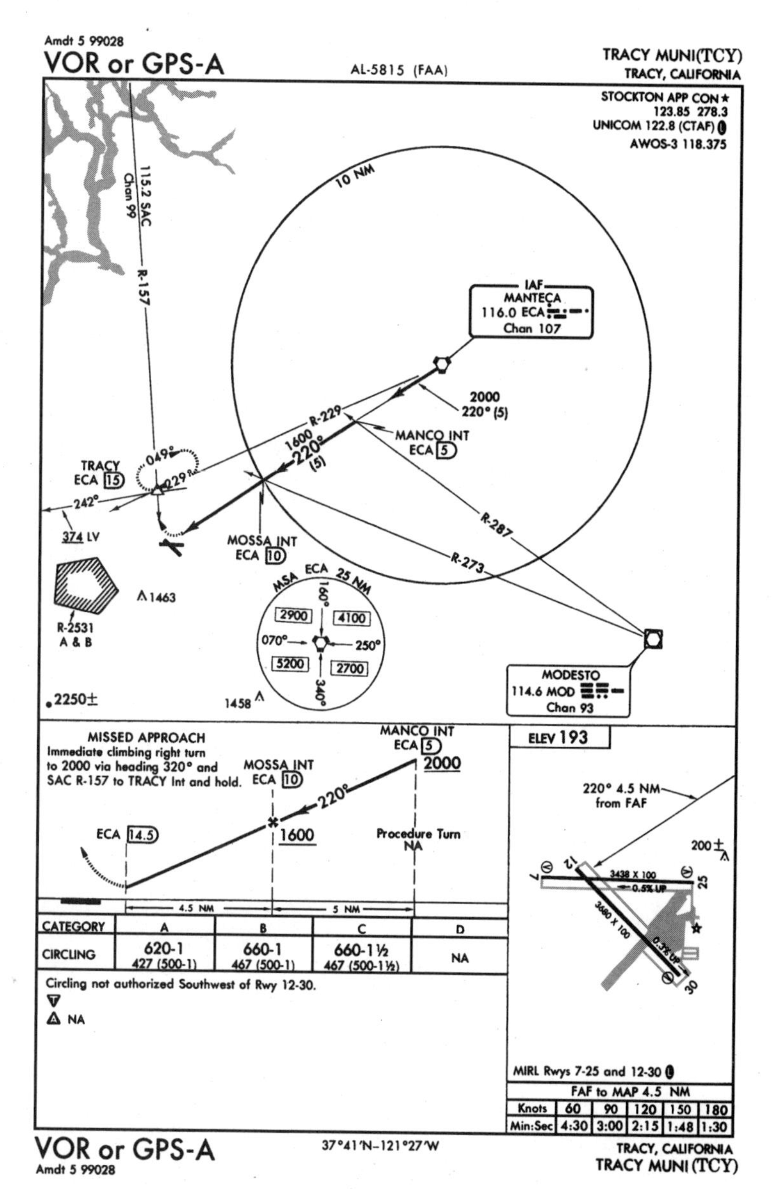

CATEGORY	A	B	C	D
CIRCLING	620-1 427 (500-1)	660-1 467 (500-1)	660-1½ 467 (500-1½)	NA

Circling not authorized Southwest of Rwy 12-30.

▼

▲ NA

FAF to MAP 4.5 NM					
Knots	60	90	120	150	180
Min:Sec	4:30	3:00	2:15	1:48	1:30

7-8 *During the flight planning stage a pilot needs to review all destination procedures; this may be as simple as one IAP or as complex as more than a dozen.*

Some pilots think they must file to the final approach fix—for example, "...ECA MANCO MOSSA TCY." However, because of computer storage capability, not all intermediate and final approach fixes can be stored. Pilots using DUAT will receive rejection messages indicating "fix not stored" or "invalid entrance/exit for airway." File to an IAF or a fix with a published transition to the approach.

From the communications data we see that Stockton approach is the controlling agency, but operates only part time (*). Where would we find the controlling agency and frequencies during the period Stockton is closed? This information is contained in the *Airport/Facility Directory*. Airport communications consist only of UNICOM, which is the CTAF. Runway lights are standard PCL, as indicated by the negative (white-on-black) L following CTAF. Finally, the communications data tell us that the Tracy Airport has an automated weather observation system (AWOS-3) on 118.375 MHz.

At this point it might be appropriate to go directly to the landing minimums sections. Let's assume our aircraft is in category B. The first thing to notice is that this approach is circling only. Category B MDA is 660 ft, visibility 1 sm. The note is extremely important. It states that: "Circling not authorized Southwest of Rwy 12-30." Why? There are obstruction and high terrain in this quadrant. While we're here note that there are nonstandard takeoff or obstacle clearance procedures for this airport; the airport is not authorized to be used as an alternate.

In what class of airspace is the Tracy Airport located? We can't tell from instrument charts. The sectional or A/FD reveals that Tracy is situated in Class G airspace, with the floor of Class E at 700 ft AGL. From the airport sketch we determine that field elevation is 193 ft. With an MDA of 660 ft, that puts the aircraft 467 ft above the airport. The MDA puts the aircraft in Class G airspace. Is it possible for VFR aircraft to be operating at Tracy clear of clouds with 1 mile visibility? Yes! This is not an isolated case. Many uncontrolled airports are located in Class G airspace with published instrument approach procedures.

The next questions is: What are the VFR traffic patterns at Tracy? Again, this cannot be determined from instrument charts. From the sectional or A/FD we determine that all patterns are left hand. We'll see what effect this has on our operation shortly.

From the plan and profile views we determine that the procedure course is made up of the ECA 220° radial. MANCO is an IF that can be established by using either the Modesto (MOD) R-287 or the ECA 5 DME; MOSSA by the MOD R-273 or the ECA 10 DME. The MAP can be established by using time from the FAF or the ECA 14.5 DME. In the profile view all of these points are shown as dashed vertical lines, indicating they are intersections.

What is the minimum equipment required to execute the Tracy VOR approach to minimums? Since there are no notes about dual VOR, DME, or ADF, a single VOR receiver is the minimum equipment needed for this approach. This will require some fast finger work to retune the VOR receiver to determine MANCO and MOSSA, then return to the approach course. Without DME how do we determine the MAP? We use time. To facilitate this, an FAF-to-MAP table, based on ground speed, is located below the airport sketch. With an approach ground speed of approximately 100 knots, this converts to a time factor of 2 minutes and 45 seconds.

Let's review the minimum safe altitudes for this approach. Note that the MSA is within a 25-nm radius of the Manteca ECA VOR, not the FAF or the airport. The arrows on the distance circle identify the sectors. For example, the southwest sector, from 160° magnetic (340°) to 250° magnetic (070°), within 25 nm of the ECA VOR is 5200 ft MSL.

As mentioned earlier the execution of an IAP results in one of two maneuvers: a landing or a missed approach. Let's take the landing first, then discuss the missed approach.

From the AWOS we determine that the wind is out of the northwest, favoring runway 30. We may or may not be able to confirm this on the CTAF. We break out of the clouds, with required visibility, just prior to the MAP. We have determined that the VFR traffic pattern for runway 30 is left. What type of pattern and entry will we make? We must enter a right downwind or right base for runway 30. Why? We are prohibited from circling southwest of runway 12-30. I think you can see the additional risk involved with this operation.

Now for the second outcome, the missed approach. The missed approach states: "Immediate climbing right turn to 2000 via heading 320° and SAC R-157 to TRACY Int and hold." Not too bad. At the missed approach point we initiate a climbing right turn via heading 320°, intercept the SAC 157° radial to the TRACY intersection and hold as published, and climb to 2000 ft. Easier said than done. The distance from the MAP to TRACY is not very far and typically happens very fast. This can be very challenging with a single VOR receiver. Notice that TRACY can be established using the 242° bearing to the LV NDB. Why is this even on the chart? Well, an aircraft with a single VOR receiver and an ADF could establish this intersection without switching the VOR back and forth from SAC to ECA.

It's winter and the wind is strong out of the southeast, favoring runway 12. From the missed approach point we would enter a left downwind or base, this time using the published pattern. We turn final and reenter the clouds. What do we do now? The published missed instructs us to make a climbing right turn. Is that our first action? No. We want to stay over the missed approach protected area, which is the runway complex, and avoid the area southwest of runway 12-30. We would initiate a climbing left turn to a heading of 320° and complete the missed approach procedure as published.

Figure 7-9 shows the Ukiah Municipal Airport, Ukiah, Calif., LOC RWY 15 approach. Notice that the chart uses the new format, with terrain depicted. A note advised that without the local altimeter setting the procedure is not authorized. Why would this be the case? The altimeter setting corrects for both nonstandard pressure and temperature at field elevation. If the atmosphere above a station has a standard lapse rate (which it rarely does), indicated and true altitude will be the same. For instrument approaches we want indicated and true altitude to be as close as possible during the final approach phase, which a current altimeter setting provides.

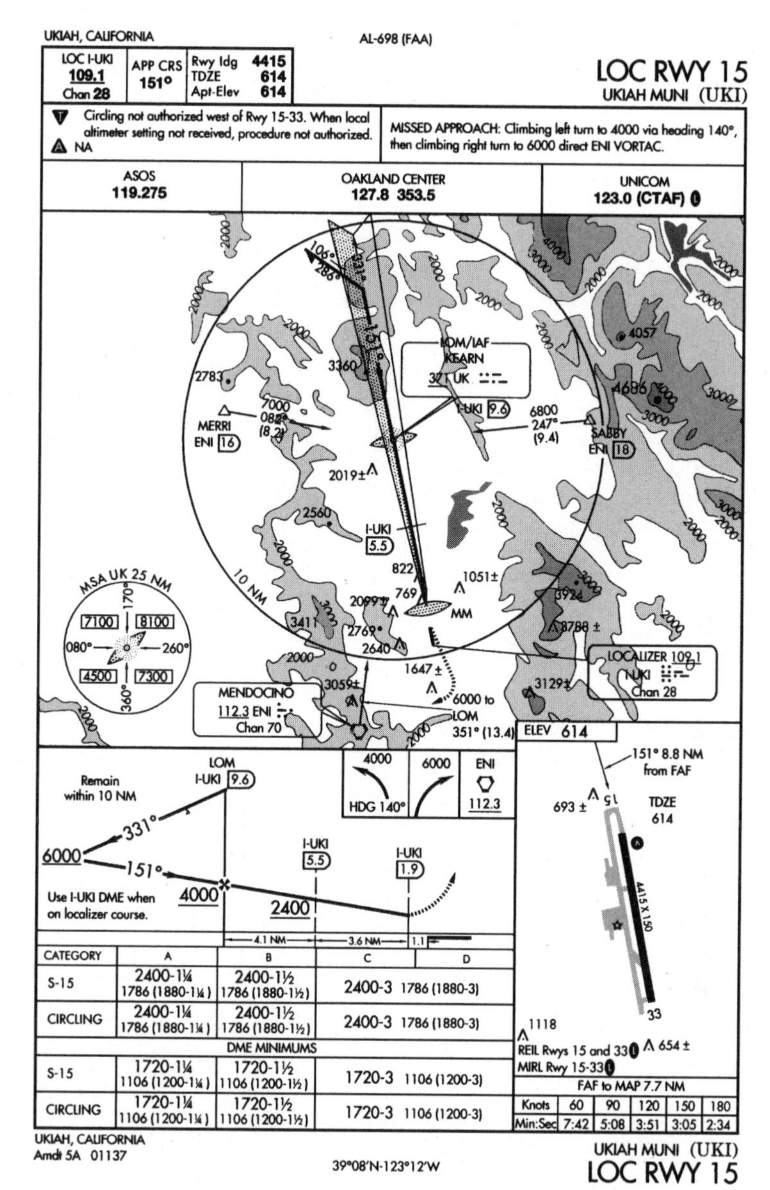

CATEGORY	A	B	C	D
S-15	2400-1¼ 1786 (1880-1¼)	2400-1½ 1786 (1880-1½)	2400-3 1786 (1880-3)	
CIRCLING	2400-1¼ 1786 (1880-1¼)	2400-1½ 1786 (1880-1½)	2400-3 1786 (1880-3)	
DME MINIMUMS				
S-15	1720-1¼ 1106 (1200-1¼)	1720-1½ 1106 (1200-1½)	1720-3 1106 (1200-3)	
CIRCLING	1720-1¼ 1106 (1200-1¼)	1720-1½ 1106 (1200-1½)	1720-3 1106 (1200-3)	

FAF to MAP 7.7 NM

Knots	60	90	120	150	180
Min:Sec	7:42	5:08	3:51	3:05	2:34

7-9 *Localizer, simplified directional facility (SDF), and localizer-type direction aid (LDA) approaches have no electronic glide slope.*

Simplified directional facility (SDF) and localizer-type direction aid (LDA) are similar to localizer approaches. None have an electronic glide slope. SDF and LDA approaches are not necessarily aligned with a runway, or may have circling minimums only. They are all flown using the same technique.

Ukiah has only one IAF-KEARN. However, three feeder routes are depicted: MERRI, SABBY, and ENI. Therefore, a pilot could fly the enroute structure to any one of these three fixes, then in the flight plan file direct UKI. How is a pilot going to navigate from MERRI or SABBY to KEARN? Although not required for the approach, an ADF receiver is required for these transitions. Otherwise, a pilot would have to fly to ENI and ENI 351° radial to KEARN. Without an ADF, a marker beacon receiver or DME would be necessary to establish KEARN. The filing of KEARN in the flight plan is not necessary, and, if fact, may not be stored in ATC computer systems. The point is that should radio communications be lost, a pilot needs published transitions to the approach procedure.

A note in the profile view instructs pilots to use the I-UKI DME when on the localizer course. This is very important. Aircraft have been lost on this approach because the pilot had the DME set to Mendocino. Although DME is not required for this approach, the use of DME allows lower minimums, which are contained in a separate section of the landing minimums data.

The missed approach procedure is pictorially depicted in the plan view.

Figure 7-10 shows the Livermore ILS RWY 25R approach. Notes state that ADF is required, circling not authorized north of runway 7L-25R, and autopilot coupled approaches not authorized below 1700. A review of the missed approach procedure shows why the ADF is required; it is needed for course guidance on the missed. TRACY and REIGA are IAFs, and a transition is provided from ALTAM. Note the lightning bold symbol indicating glide slope intercept at 2800 ft outside of the outer marker.

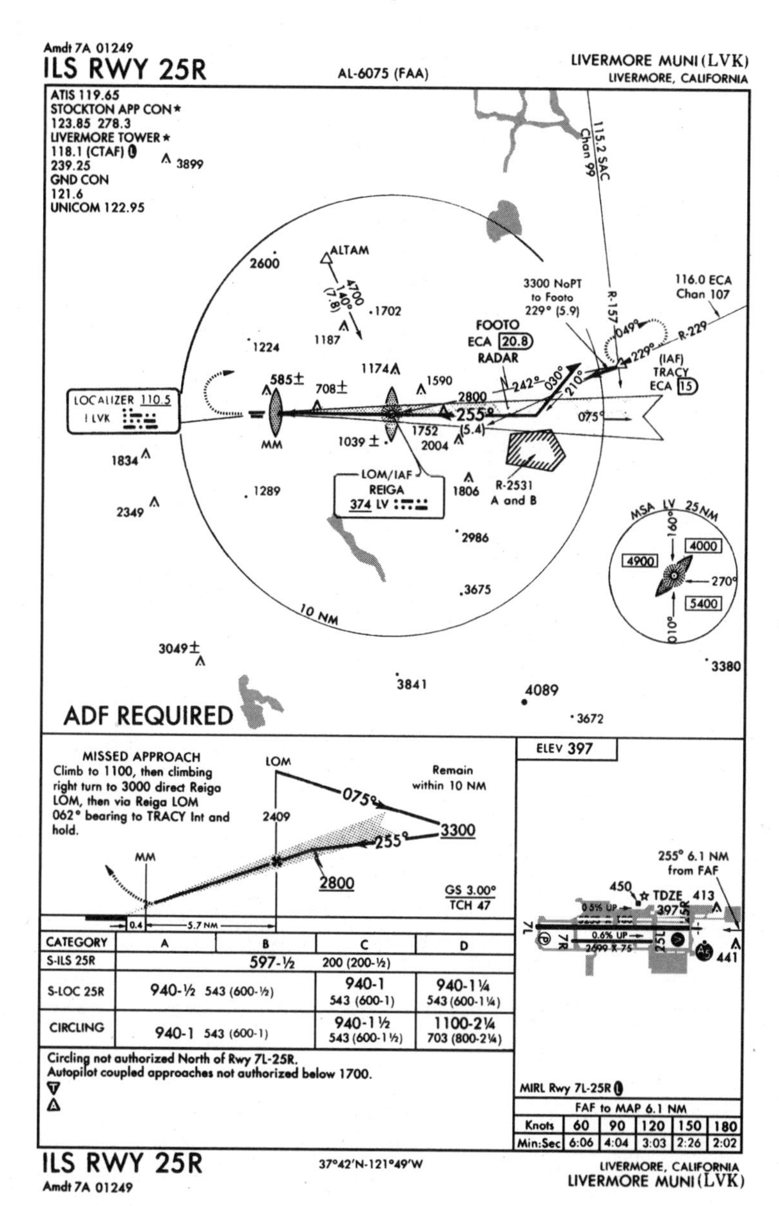

CATEGORY	A	B	C	D
S-ILS 25R	597-½		200 (200-½)	
S-LOC 25R	940-½ 543 (600-½)		940-1 543 (600-1)	940-1¼ 543 (600-1¼)
CIRCLING	940-1 543 (600-1)		940-1½ 543 (600-1½)	1100-2¼ 703 (800-2¼)

FAF to MAP 6.1 NM					
Knots	60	90	120	150	180
Min:Sec	6:06	4:04	3:03	2:26	2:02

7-10 *A review of notes is extremely important. Notes state ADF is required; a review of the missed approach procedure shows ADF is needed for course guidance on the missed approach.*

The approach has straight-in ILS and LOC, and circling minimums. Could a pilot use this approach for a landing on 25L? Yes, by applying circling minimums. What are the minimums based on the following clearance: "CLEARED FOR AN ILS 25R APPROACH, CIRCLE TO LAND RUNWAY 7 LEFT." That's simple, circling minimums for category A, B, and C are MDA 940 ft and we're required to circle to the south. In the calm atmosphere of our favorite easy chair the answer seem obvious. But, in a busy terminal area, at night, with turbulence, a single pilot operation, and the tower closed, the subtleties of this clearance could be missed.

Prior to the installation of the AWOS, this procedure was not authorized when the control tower was closed. This was because of the lack of an altimeter setting. The notes section indicates that this airport has nonstandard takeoff or obstacle clearance procedures, and nonstandard alternate minimums. Recall our discussion of the LIVERMORE ONE departure. A check of the TPP shows that this airport is not authorized as an alternate when the control tower is closed. Why? The tower monitors the approach aids. When the tower is closed the aids are unmonitored, and therefore this airport cannot be used as an alternate.

Case Study

The weather at Livermore was wind 060° at 20 knots, visibility 1, ceiling 400 ft overcast. The ATIS was advertising ILS approach runway 25R in use, landing runway 25R, departing runway 7L. Why? The airport was below circling minimums; therefore, the only approach below the overcast was the ILS to runway 25R. Our ground speed on the approach in the Cessna 172 was 110 knots. The landing was extremely exciting with the tailwind and we used quite a bit of runway. Should a pilot of an MD-80 jet airliner attempt this approach? Maybe not. The runway is only 5253 ft. It would certainly require a careful look at landing distance required. Oh yes, we do have an MD-80 based at Livermore.

Do you think there would be extensive departure and arrival delays for Livermore under these conditions. Absolutely! Recall our previous discussion of head-on traffic.

Figure 7-11 contains the Oakland ILS RWY 29 (CAT II) approach. In addition to outer and middle markers, this procedure has an inner marker approximately, 1/8 mile from the runway. The main difference between category I and category II procedures is the landing minima and the fact that the approach requires a full ILS to a specific runway—no localizer or circling minima. The minima are a radar altimeter DH of 155 ft or 105 ft, depending upon the aircrew and aircraft certification.

In addition to categories I and II, there is a category III ILS approach. Category III is divided into three sections: CAT IIIa, CAT IIIb, and CAT IIIc. There is no DH because CAT III consists of an automated landing to the runway. The difference is required visibility. CAT IIIa requires an RVR of 700 ft, IIIb 600 ft; with IIIc zero-zero landings are authorized. The problem then becomes: How do you get to the terminal?

Unlike a localizer course that has an accuracy of plus or minus one degree, an NDB bearing course is considered to have an error of 5°. This error is cumulative, because of equipment—airborne receiver and heading indicator error—and pilotage, which is assumed to be controlled within normal tolerance. How about compass deviation? One consequence is that NDB minimums are usually higher than minimums for procedures using other course guidance. This is illustrated by a complaint from a large flight school that an NDB was out of tolerance.

Case Study

Pilots at the FAF were complaining of being a mile off course. The NDB was collocated at the ILS middle marker, 7.6 nm from the FAF. By calculating course guidance accuracy and adding as little as 2 1/2° of compass deviation, 1-mile error at the FAF is not unreasonable. Pilots need to be aware of these parameters.

Let's say that Tracy is our destination and we've lost radio communications. We would proceed at the last assigned altitude or the MEA, whichever is higher. We've planned ahead and filed a route that will take us to the ECA VOR. If our clearance limit is the airport, we commence the approach as

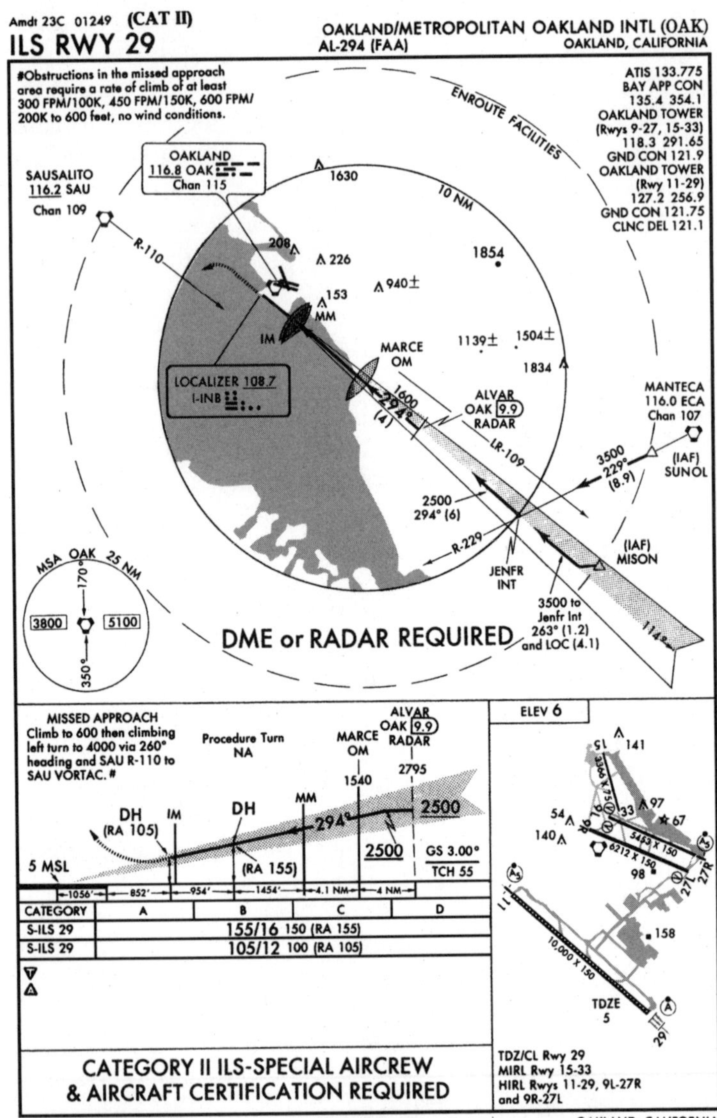

7-11 *A category II approach requires special certification for aircrew and aircraft; converging approaches must meet special criteria.*

soon a we reach ECA, regardless of the time over the fix. If ATC has issued an expect approach clearance time, we will begin a descent and approach as close as possible to that time. If we were cleared short, for example to the ECA VOR without an expect further clearance or approach time, we would commence the approach at our flight plan filed estimated time of arrival.

Normally, ATC will issue EFC times and route when holding aircraft short of the destination. For example, in this case we might receive a clearance to "hold, expect further clearance at zero five one five, maintain 6000." At zero five one five we would depart ECA and commence descent and approach. One final thought, if communications are lost don't forget to monitor VOR voice communication because ATC will attempt to communicate on these frequencies.

RNAV approach procedures

VORTAC RNAV approach charts were developed in 1971. LORAN RNAV approach charts were introduced in 1990.

On July 21, 1994, the first stand-alone GPS approach was issued, the Frederick, Md., GPS RWY 5 approach. A stand-alone procedure can be executed only with an IFR-approved GPS receiver. Chart symbology is standard. Since waypoints are stored in the GPS receiver database, waypoints are not defined by radial/distance or latitude/longitude coordinates. Even though the RNAV systems have an extensive database, not all navigation information is provided. They are not a substitute for current aeronautical charts.

The GPS approach overlay program is an authorization for pilots to use GPS, under IFR, for flying designated non-precision-instrument approach procedures—except LOC, LDA, and SDF. Aircraft using GPS under IFR must be equipped with an approved and operational alternate means of navigation appropriate to the flight. Only approaches contained in the current onboard navigation database are authorized. Even though the database contains an approach, aircraft equipment may not be certified for all procedures.

GPS avionics approved for terminal IFR operations may be used in lieu of ADF and DME. If an alternate is required, a non-GPS approach must exist at the alternate airport. If ADF or DME are required at the alternate, the aircraft must be equipped with ADF or DME as appropriate. For example, the Livermore ILS requires ADF. Therefore, an aircraft with ILS and GPS, but without an ADF receiver cannot use Livermore as an alternate.

Unnamed stepdown fixes in the final approach segment will not be coded in the waypoint sequence of the aircraft's GPS navigation database. Without vertical guidance, pilots must use charted along track distance (ATD) on the GPS navigational system to identify stepdown fixes. Nor are visual descent points included in the sequence of waypoints.

RNAV instrument approach procedures include all types of approaches using area navigation, whether ground or satellite based. RNAV minimums are dependent on navigation equipment capability, as stated in the applicable airplane flight manual (AFM) and as outlined below. Figure 7-12 contains an example of an RNAV landing minimum data section.

Global navigation system (GNS) landing system (GLS) provides precision navigation guidance for exact alignment and descent of aircraft on approach to a runway. GLS must have wide area augmentation system (WAAS) equipment approved for precision approach (PA). PA indicates that the runway environment meets precision approach requirements. If the GLS minimum line does not contain PA, then the runway environment does not support precision requirements. Minimums are indicated by a decision altitude (DA) and visibility in statute miles or RVR (DA 1382 ft, RVR 2400). (DA is an ICAO term for a specified altitude in a precision approach, where a missed approach must be initiated if required.)

Lateral navigation/vertical navigation (LNAV/VNAV) must have WAAS equipment approved for precision approach, or (required navigational performance) RNP-0.3 system based on GPS or DME/DME, with an IFR approach approved Baro-VNAV system. Other RNAV approach systems require special approval.

CATEGORY	A	B	C	D
GLS PA DA	1382/24 200 (200-½)			
LNAV/DA VNAV	1500/24 318 (400-½)			1500/50 318 (400-1)
LNAV MDA	1700-24 518 (600-½)		1700-50 518 (600-1)	1740-2 561 (600-2)

7-12 *RNAV instrument approach procedures include all types of approaches using area navigation, whether ground or satellite based.*

Lateral navigation (LNAV) must have IFR approach approved WAAS, GPS, GPS-based FMS systems, or RNP-0.3 systems based on GPS or DME/DME. (If you don't know what this means, then you don't have it on your aircraft. The bottom line: If the aircraft is equipped with an IFR approved GPS, with a current database, LNAV and circling minimums apply.)

The terminal arrival area (TAA) provides a transition from the enroute structure to the terminal environment for aircraft equipped with FMS or GPS. A TAA will not be found on all RNAV procedures. The TAA replaces the MSA for that approach procedure.

Refer to Fig. 7-13. The TAA provides a basic or modified T that incorporates from one to three IAFs, an intermediate fix (IF) that also serves as an IAF, an FAF, a MAP, and a missed approach holding fix (MAHF). A standard holding pattern may be provided at the center IAF, and, if necessary, a course.

The standard TAA consists of three areas: the straight-in, left base, and right base. Lateral boundaries are defined by magnetic courses *to* the IF (IAF). Areas may be further divided into stepdown fixes defined by courses and arcs, with minimum altitudes charted. Pilots entering the TAA and cleared by ATC are expected to proceed directly to the IAF associated with that portion of the TAA at the altitude depicted, unless otherwise authorized by ATC.

Figure 7-14 contains the San Jose International, San Jose, Calif., GPS RWY 30L approach. From the procedure identification

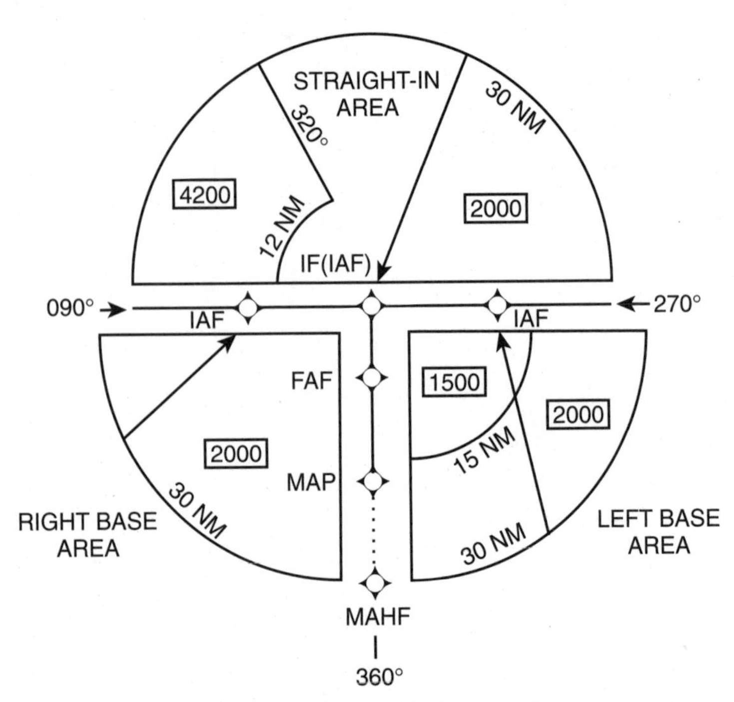

7-13 *The terminal arrival area (TAA) provides a transition from the enroute structure to the terminal environment for aircraft equipped with FMS or GPS.*

this approach is authorized only for RNAV aircraft equipped with GPS. A TAA is not appropriate, and therefore omitted, from this procedure. At the center top of the procedure, in the briefing strip, is the type of ALS—a MALSR, coded A5. Since the code is in negative format, the MALSR is pilot controlled. Notice in the plan and profile view that there are two ATD stepdown fixes. One fix is 5 nm to IRONN and the other is the VDP, 1.6 nm to RW30L—the missed approach point. Since this procedure does not have a TAA, MSA is depicted. The MSA is based on the RW30L waypoint.

The landing minima data shows sidestep minimums for both runway 30L and 29. Interestingly, this approach is not authorized for circling. From the airport sketch we see that all runways are served by a PAPI, the small circle with a P. This can be verified in the *Airport/Facility Directory*.

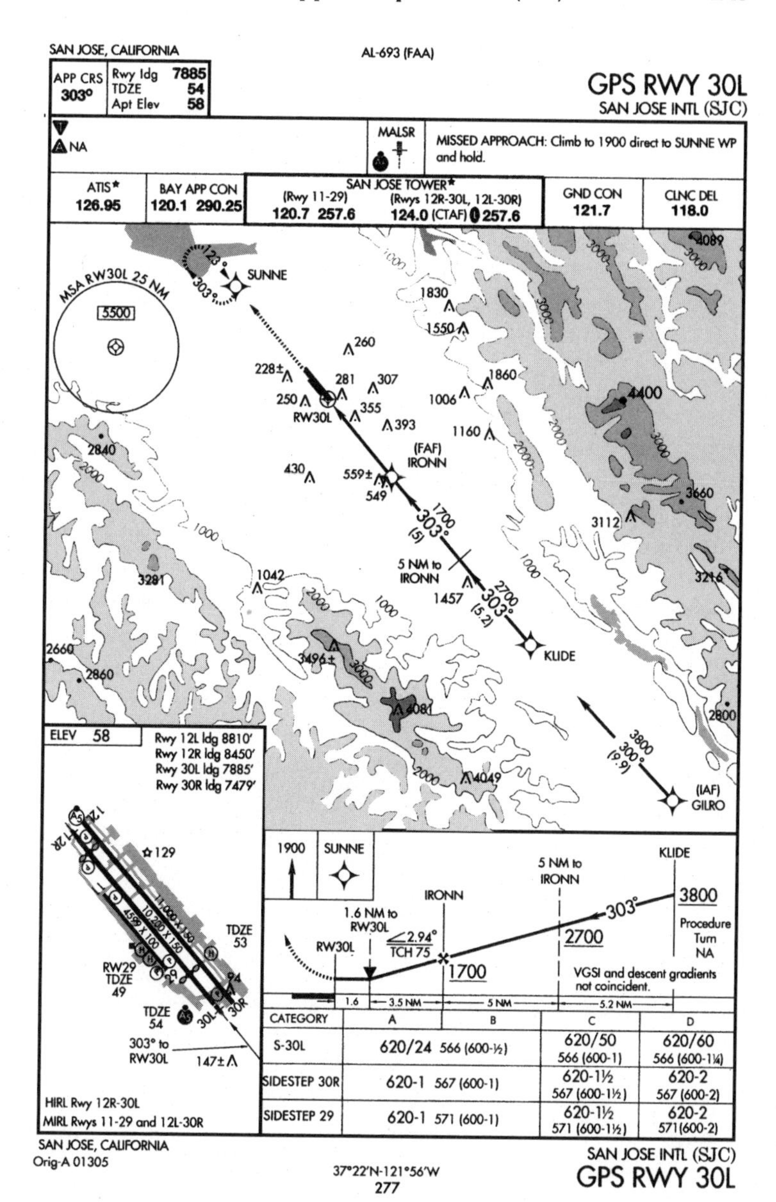

7-14 *From the procedure identification, this approach is authorized only for RNAV aircraft equipped with IFR-approved GPS equipment.*

Other approach charts

Howie Keefe's Air Chart Service provides an approach chart service. Air Chart Service uses NACO plates, offered in a gum-bound as well as loose-leaf format. Advantages advertised over Jeppesen charts are low cost and easy to update.

E. B. Jeppesen first published his *Airway Manual* in 1934. Jeppesen services and products have become worldwide since then. Jeppesen charts are printed on 5½ × 8½-in sheets; oversized charts are folded to the standard size. The most often heard disadvantage of the individual Jeppesen charts, especially the approach plates, is the constant requirement to update the *Airway Manual*. (Some pilots prefer NACO charts because they are simply thrown away and entirely replaced.)

Jeppesen provides IAP chart coverage for the world. Fifteen coverage options are available for the United States, ranging from the entire country to sections similar to individual volumes of the TPP. In 1994 Jeppesen introduced a color-contour in the plan view for each approach with terrain higher than 4000 ft above the runway. Canada, Alaska, and the Aleutians have three coverage areas; therefore, pilots need to subscribe only to those areas where they do most of their flying. Jeppesen provides special trip coverage on a one-time-sale basis.

8

Publications

Supplemental aeronautical information was published in the *Domestic Air News,* until 1929 when it was replaced by *Air Commerce Bulletins.* These publications contained official aviation information assembled and distributed by the Department of Commerce. In 1932, the free bulletin series described airports, intermediate landing fields, and meteorological conditions in the various states, along with the low-frequency radio ranges for air navigation. By 1940, bulletins contained notices to airmen, air navigation radio aids, danger areas in air navigation, and a directory of airports. These were the forerunners of today's *Airport/Facility Directory* and *Notice to Airmen* publication.

The Aeronautical Chart and Information Center introduced the Flight Information Publication—Planning (FLIP) in 1958 to further eliminate nonessential material. This publication contained charts and textual data necessary for flight planning. FLIP, and other publications of this type, provide supplemental information that cannot be printed on the chart because of space constraints. The *Airman's Information Manual* (AIM) replaced the *Airman's Guide* in 1964. Since then the AIM has gone through various evolutionary stages, at times consisting of four documents, to its present form of *Basic Flight Information and ATC Procedures, Airport/Facility Directory,* and *Notices to Airmen.* In 1995 the *Airmen's Information Manual* was replaced with the *Aeronautical Information Manual.* The change is in name only.

Effective application of charts during preflight planning and while navigating enroute is dependent upon an understanding of supplemental publications. Simple charts published in the early days of aviation naturally evolved hand in hand with the complexities of flying, and before too long charts could no longer reasonably depict all the data. Charted information changed more rapidly than it was possible to update, reprint, and distribute in a cost-efficient manner. Most aeronautical publications are merely extensions of aeronautical charts.

Aeronautical publications are most often produced and supplemented by the same agency that publishes the respective charts; NACO charts are supplemented by the *Airport/Facility Directory* (A/FD) published by FAA, and the NOTAM publication and NOTAM system administered by the FAA. NIMA supports its charts through flight information publications (FLIPs) and the aeronautical chart updating manual (CHUM). Canada and other countries that produce charts have similar supplementary publications, such as the *Canada Flight Supplement*.

This chapter focuses on publications that supplement NACO visual and instrument aeronautical charts: the A/FD, Alaska supplement, Pacific chart supplement, and the *Notice to Airmen* (NOTAM) publication. These are illustrated in Fig. 8-1. Chapter 9 discusses the FLIP, CHUM, and other supplementary publications.

I subscribe to the volume of the A/FD for the area where I do most of my flying. It's in my flight case and I have found it to be of immense value. Although charts provide essential data, the directory provides the details. For long trips, out of the area of coverage of my directory, I visit the FSS and use the directories that cover the route. Directories are also available on a one-time sale basis from many chart suppliers. Pilots planning long trips would be well advised to obtain the directories that cover their route. From a review of the directory, all pertinent data can be obtained and noted or logged on a flight planning form.

Airport/Facility Directory

The A/FD is published in paperback books $5^{3}/_{8} \times 8^{1}/_{4}$ in, on the standard 56-day revision cycle. The directory is an alphabetical listing of data on record with the FAA for all airports

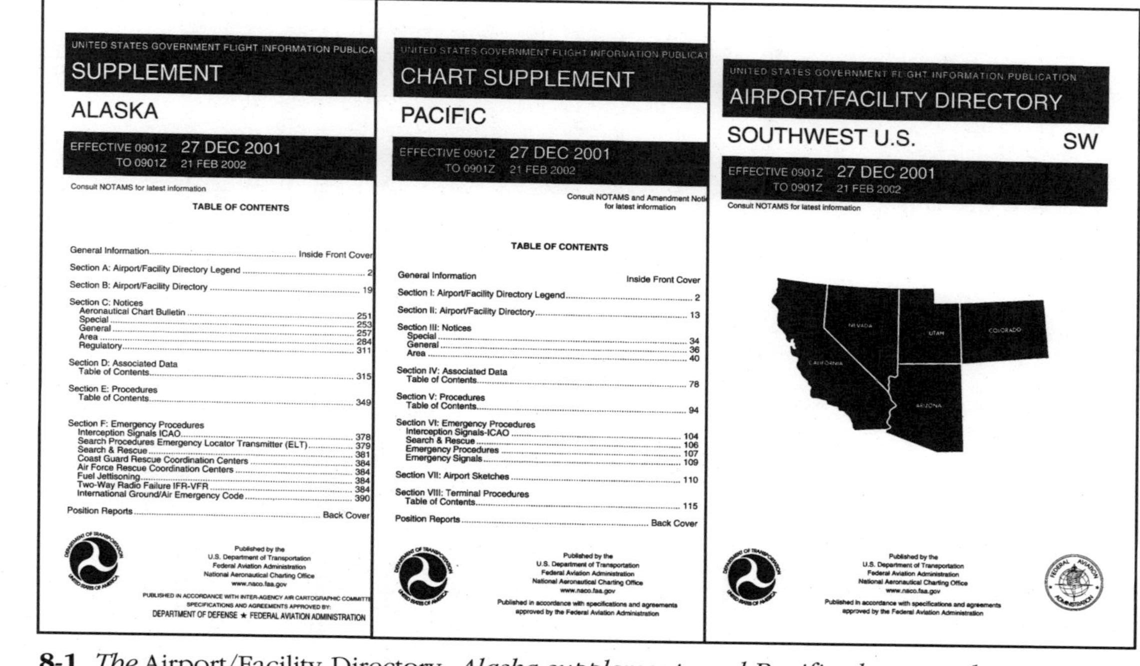

UNITED STATES GOVERNMENT FLIGHT INFORMATION PUBLICA

SUPPLEMENT

ALASKA

EFFECTIVE 0901Z 27 DEC 2001
TO 0901Z 21 FEB 2002

Consult NOTAMS for latest information

TABLE OF CONTENTS

General Information .... Inside Front Cover
Section A: Airport/Facility Directory Legend 2
Section B: Airport/Facility Directory 19
Section C: Notices
Aeronautical Chart Bulletin 251
Special 253
General 257
Area 284
Regulatory 311
Section D: Associated Data
Table of Contents 315
Section E: Procedures
Table of Contents 349
Section F: Emergency Procedures
Interception Signals ICAO 378
Search Procedures Emergency Locator Transmitter (ELT) 379
Search & Rescue 381
Coast Guard Rescue Coordination Centers 384
Air Force Rescue Coordination Centers 384
Fuel Jettisoning 384
Two-Way Radio Failure IFR-VFR 384
International Ground/Air Emergency Code 390
Position Reports .... Back Cover

Published by the
U.S. Department of Transportation
Federal Aviation Administration
National Aeronautical Charting Office
www.naco.faa.gov

PUBLISHED IN ACCORDANCE WITH INTER-AGENCY AIR CARTOGRAPHIC COMMITTE
SPECIFICATIONS AND AGREEMENTS APPROVED BY:
DEPARTMENT OF DEFENSE ★ FEDERAL AVIATION ADMINISTRATION

UNITED STATES GOVERNMENT FLIGHT INFORMATION PUBLICAT

CHART SUPPLEMENT

PACIFIC

EFFECTIVE 0901Z 27 DEC 2001
TO 0901Z 21 FEB 2002

Consult NOTAMS and Amendment Noti
for latest information

TABLE OF CONTENTS

General Information .... Inside Front Cover
Section I: Airport/Facility Directory Legend 2
Section II: Airport/Facility Directory 13
Section III: Notices
Special 34
General 36
Area 40
Section IV: Associated Data
Table of Contents 78
Section V: Procedures
Table of Contents 94
Section VI: Emergency Procedures
Interception Signals-ICAO 104
Search & Rescue 106
Emergency Procedures 107
Emergency Signals 109
Section VII: Airport Sketches 110
Section VIII: Terminal Procedures
Table of Contents 115
Position Reports .... Back Cover

Published by the
U.S. Department of Transportation
Federal Aviation Administration
National Aeronautical Charting Office
www.naco.faa.gov

Published in accordance with specifications and agreements
approved by the Federal Aviation Administration

UNITED STATES GOVERNMENT FLIGHT INFORMATION PUBLICATION

AIRPORT/FACILITY DIRECTORY

SOUTHWEST U.S. SW

EFFECTIVE 0901Z 27 DEC 2001
TO 0901Z 21 FEB 2002

Consult NOTAMS for latest information

Published by the
U.S. Department of Transportation
Federal Aviation Administration
National Aeronautical Charting Office
www.naco.faa.gov

Published in accordance with specifications and agreements
approved by the Federal Aviation Administration

8-1 *The* Airport/Facility Directory, *Alaska supplement, and Pacific chart supplement support NACO visual and instrument charts for their respective regions.*

that are open to the public, associated terminal control facilities, air route traffic control centers (ARTCCs), and radio aids to navigation within the contiguous United States, Puerto Rico, and the Virgin Islands. Radio aids and airports are listed alphabetically. Airports and associated cities are cross-referenced when necessary. The directory directly supports visual charts through the airport listing and aeronautical chart bulletin. Instrument charts are supported through NAVAID restrictions, and detailed airport services and information not available on charts.

VFR airport sketches are being added to the A/FD. Sketches depict more visual information than found on airport sketches in the TPP. Added visual features include:

- Buildings
- Airport road pattern
- Vegetation features
- Symbology used for wind socks and rotating beacons

The sketches will focus on VFR airports with single configuration hard surface runways at least 3000 ft and IAP procedures.

Beginning July 12, 2001, airport diagrams for selected airports certified for air carrier operations will be added as a separate section to the A/FD. Diagrams will be listed in order by associated city and airport name. Airport diagrams (5 × 8 in) will depict current runway and taxiway configurations and will assist both VFR and IFR pilots in ground taxi operations at large, complex metropolitan airports. Airport diagrams are the same full-page charts as those published in the TPP, and will be in addition to the VFR sketches that are already contained in the A/FD. Airport diagrams are also available on the National Aeronautical Charting Office Internet site: www.naco.faa.gov.

The A/FD is divided into seven booklets; coverage is depicted in Fig. 8-2. Each directory contains the following:

- General information
- Abbreviations
- Legend, A/FD

- A/FD
- Seaplane landing areas
- Notices
- Land and hold-short operations
- Simultaneous operations on intersecting runways
- FAA and National Weather Service telephone numbers
- Air route traffic control centers
- FSDO addresses and telephone numbers
- Routes/waypoints
- VOR receiver checkpoints
- Parachute-jumping areas
- Aeronautical chart bulletin
- Tower enroute control
- Airport diagrams
- National Weather Service upper air observing stations
- Enroute flight advisory service

The inside front cover of the directory provides general information. This consists of information about corrections,

8-2 *The seven booklets of the* Airport/Facility Directory *cover the United States, including Puerto Rico and the Virgin Islands.*

comments, and procurement of the directory, with addresses and telephone numbers. General information contains the directory publishing schedule, along with airport and airspace information publication cutoff dates.

Next is a table of contents and list of abbreviations commonly used in the directory. Other abbreviations are contained in the legend and not duplicated in the list. (Abbreviations are also contained in App. A, NOTAM, chart contractions.) This section is followed by the directory legend.

The A/FD lists public-use airports and NAVAIDs that are part of the National Airspace System (NAS). These are listed alphabetically by city or facility name, within the state, and cross-referenced when necessary. This section is followed by a listing of public-use seaplane landing areas.

The notices section contains additional information within the coverage of the directory. Other data in this section consist of advance flight plan filing requirements, special flight procedures, temporary closure of facilities, and other general information. It advertises laser light demonstrations, controlled firing areas, flight test operations, border crossing procedures, and aerobatic operations. Keep in mind that information contained in the publications will normally not be provided during an FSS or DUAT briefing.

A listing of land and hold-short operations (LAHSO) and simultaneous operations on intersecting runways (SOIR) is provided. The *Aeronautical Information Manual* contains specific details on hold-short operations and markings, and pilot and controller responsibilities.

Telephone numbers are provided for FAA and NWS facilities within the directory area of coverage, listed alphabetically within the state. The availability of special services is noted—for example, recorded aviation weather and fast-file flight plan filing services. Flight standards district office (FSDO) addresses and telephone numbers are listed.

Air route traffic control center sector frequencies are provided. These are listed alphabetically by location and altitude stratum (low or high), within the individual ARTCC, for the area of coverage of the directory.

Preferred IFR routes are listed in the routes/waypoints section. IFR preferred routes have been established to help pilots in planning their routes, to minimize route changes during flight, and aid in the efficient, orderly management of air traffic. VFR waypoint names consist of five letters beginning with VP. Stand-alone VFR waypoints are portrayed on VFR charts using the same four-point star symbol used on instrument charts. VFR waypoints collocated with visual check points are portrayed with a checkpoint flag on aeronautical charts. The VFR waypoint name is shown in parentheses adjacent to the visual checkpoint name. The listing consists of waypoint identification (VPLQM), collocated VFR checkpoint if applicable (QUEEN MARY), and location (N33°44.43′/W118°11.37′).

Approved VOR receiver checkpoints and VOR test facilities (VOT) are listed alphabetically, within the states. Type of check, ground or airborne, and checkpoint description are provided. VOT facilities are listed separately.

Parachute-jumping areas, depicted by a small parachute symbol on sectional and TAC charts, are tabulated alphabetically, within the states. This is where a pilot would obtain details on charted parachute-jumping areas. Unless otherwise indicated, all activities are conducted during daylight hours and in VFR weather conditions. This section also outlines procedures for parachute-jumping areas to qualify for inclusion on charts.

The aeronautical chart bulletin provides major changes in aeronautical information that have occurred since the last publication date of each sectional, terminal area, and helicopter chart. Additionally, users of world aeronautical and United States Gulf Coast VFR aeronautical charts should make appropriate revisions to their charts from this bulletin.

Tower enroute control (TEC) refers to IFR operations conducted entirely within approach control airspace. These listings are similar to preferred routes. Pilots can use this section to determine preferred routes, altitudes, and type of operations allowed (jet, turboprops, nonjet, and the like).

In support of the FAA's runway incursion program, selected towered airport diagrams have been published in the "Airport Diagrams" section of the A/FD. Some text data published under the individual airport in the front portion of the A/FD may be more current than the data published on the airport diagrams. The airport diagrams are updated only when significant changes occur.

The page opposite the inside back cover of the directory contains a graphic of National Weather Service (NWS) upper air observing stations and weather radars. These are scheduled balloon releases (1100 UTC—coordinated universal time—and 2300 UTC); therefore, pilots cannot expect to be notified by NOTAM, unless the release is delayed beyond 1130 UTC or 2330 UTC. NOTAMs are issued for unscheduled balloon releases.

The inside back cover of the directory provides the location and frequency of flight watch outlets within the area served by the volume. All low-altitude flight watch outlets are on the common frequency of 122.0 MHz; therefore, only individual flight watch control station high-altitude frequencies are listed. Each center's airspace is served by a discrete high-altitude flight watch frequency.

Listings in the airport/facility portion of each directory merely indicate the airport operator's willingness to accommodate transient aircraft, and does not represent that the facility conforms with any federal or local standards, or that it has been approved for use by the general public. Information on obstructions is taken from reports submitted to the FAA. This information has not been verified in all cases. Pilots are cautioned that obstructions not indicated in this tabulation, or on charts, might exist that can create a hazard to flight operations. Detailed specifics concerning services and facilities in the directory are contained in the *Aeronautical Information Manual.*

Directory legend

Figure 8-3 is an example entry in the directory. Features include:

- City/airport name
- Location identifier
- Airport location

- Time conversion
- Geographic position of airport
- Charts on which the facility is located
- Airport sketch
- Elevation
- Rotating light beacon
- Servicing available
- Fuel available
- Oxygen available
- Traffic pattern altitude

OAKLAND

METROPOLITAN OAKLAND INTL (OAK) 4S UTC-8(-7DT) N37°43.28'WI22°13.24' **SAN FRANCISCO**

06 B S4 FUEL 100LL, JET A OX 1,2,3,4 TPA-See Remarks LRA ARFF Indx D **H-2A,L-2F,A**
RWY 11-29: H10000X150 (ASPH-PFC) S-200, D-200, DT-400, DDT-900 HIRL CL **IAP, AD**
RWY 11: MALSR. Rgt tfc. **RWY 29:** ALSF2 TDZ.
RWY 09R-27L: H6212XI50 (ASPH-PFC) S-75, D-200, DT-400, DDT-800 HIRL
RWY 09R: VASI(V4L)-GA 3.0 TCH 46'. Tree. **RWY 27L** VASI(V4L)-GA 3.0 TCH 55'.
RWY 09L-27R: H5453XI50 (ASPH) S-75,D-115,DT-180 HIRL
RWY 09L VASI(V4L)-GA 3.0 TCH 381. **RWY 27R:** MALSR. Building. Rgt tfc.
RWY 15-33: H3366X75 (ASPH) S-12.5, D-65, DT-100 MIRL
RWY 33: Rgt tfc.

AIRPORT REMARKS: Attended continuously. Fee Rwy 11-29 and tiedown. Birds on and in vicinity of arpt. Rwy 09L-27R and Rwy 15-33 CLOSED to air carrier acft, except air carrier acft may use Rwy 09L and 27R for taxiing. Rwy 09L-27R and Rwy 09R-27L CLOSED to 4 engine wide body acft except Rwy 09R-27L operations avbl PPR call operations supervisor 510-577-4067. All turbo-jet/fan acft, all 4-engine acft and turbo-prop acft with certificated gross weight over 12,500 pounds are prohibited from tkf Rwys 27R/27L or ldg Rwy 09L and Rwy 09R. Preferential rwy use program in effect 0600-1400Z‡: All acft preferred north fld arrive Rwys 27R/27L or Rwy 33; all acft preferred north fid dep Rwys 09R/09L or Rwy 15. It these rwys unacceptable for safety or ATC instructions then Rwy 11-29 must be used. Prohibitions not applicable in emerg or whenever Rwy 11-29 is closed due to maintenance, construction or safety. For noise abatement information ctc noise abatement office at 510-577-4276. 400' blast pad Rwy 29 and 500' blast pad Rwy 11. Rwy 29 and Rwy 27L distance remaining signs left side. Acft with experimental or limited certification having over 1,000 horsepower or 4,000 pounds are restricted to Rwy 11-29. Rwy 09R-27L FAA gross weight strength DC 10-10 350,000 pounds, DC 10-30 450,000 pounds, L-1011 350,000 pounds. Rwy 11-29 FAA gross weight strength DC 10-10 600,000 pounds, DC 10-30 700,000 pounds, L-1011 600,000 pounds. TPA-Rwy 27L 606(600), TPA-Rwy 27R 1006(1000). Rwy 29 centerline lgts 6500'. Flight Notification Service (ADCUS) available.
WEATHER DATA SOURCES: ASOS (510) 383-9514. **HIWAS** 116.8 OAK.
COMMUNICATIONS. ATIS 128.5 (510) 635-5850 (N and S Complex) **UNICOM** 122.95
OAKLAND FSS (OAK) on arpt. 122.5 122.2. TF 1-800-WX-BRIEF. NOTAM FILE OAK.
® **BAY APP CON** 135.65 133.95 (South) 135.4 134.5 (East) 135.1 (West) 127.0 (North) 120.9 (Northwest) 120.1 (Southeast)
® **BAY DEP CON** 135.4 (East) 135.1 (West) 127.0 (North) 120.9 (Northwest)
OAKLAND TOWER 118.3 (N Complex) 127.2 (S Complex) 124.9
GND CON 121.75 (S Complex) 121.9 (N Complex) **CLNC DEL** 121.1
AIRSPACE: CLASS C svc continuous ctc **APP CON**
RADIO AIDS TO NAVIGATION: NOTAM FILE OAK.
OAKLAND (H) VORTACW 116.8 OAK Chan 115 N37°143.55'W122°13.42' at fld.10/17E. **HIWAS.**
RORAY NDB (LMM) 341 AK N37°43.28l'W122°11.65' 253° I.3 NM to fld.
ILS 108.7 I-INB Rwy29
ILS 111.9 I-AAZ Rwy11 Glideslope unusable blo 375'.
ILS 109.9 I-OAK Rwy27R LMM RORAY NDB.
COMM/NAV/WEATHER REMARKS: Emerg frequency 121.5 not avbl at twr. Emerg frequency 121.5 not avbl at FSS. Rwy 11 gild slope deviations are possible when critical areas are not required to be protected. Acft operating invof glide slope transmitter.

8-3 *The* Airport/Facility Directory *provides the detailed information that cannot be included on charts.*

- Airport of entry and landing rights airports
- FAR 139 crash, fire, rescue availability
- FAA inspection data
- Runway data
- Airport remarks
- Weather data sources
- Communications
- Airspace
- Radio aids to navigation
- Bearing and distance from nearest usable VORTAC

City/airport name. Airports and facilities are listed alphabetically by associated state and city. Where a city name is different from the airport name, the city name appears on the line above the airport name, as in Fig. 8-3—the city name is Oakland; the airport name is Metropolitan Oakland International. Airports with the same associated city name are listed alphabetically by airport name and will be separated by a dashed line. All others are separated by a solid line.

Location identifier. The official location identifier is a three- or four-character alphanumeric code assigned to the airport. These identifiers are used by ATC in lieu of the airport name for flight plans and in computer systems. It's important to distinguish between the letter O and the number 0. In Figure 8-3, the identifier for the airport is Oakland (OAK). Some of the more interesting location identifiers are VT11, 4OH4, OW15, and 84XS. The second example is "four oscar hotel four," not to be confused with "four zero hotel four." This is particularly significant for pilots obtaining weather briefings and filing flight plans with DUATS.

Airport location. Airport location is expressed as distance and direction from the center of the associated city in nautical miles and cardinal points (4 S, four nautical miles south of the city).

Time conversion. Hours of operation of all facilities are expressed in coordinated universal time (UTC) and shown as Z, or zulu, time. The directory indicates the number of hours to be subtracted from UTC to obtain local standard time and local daylight savings time [UTC-8(-7DT)], subtract 8 hours from UTC to obtain local standard time and 7 hours to obtain local daylight time). The symbol ‡ indicates that during periods of daylight savings time, effective hours will be 1 hour earlier than shown. For example in the eastern time zone tower: 1100–2300‡. This tower operates from 6 A.M. until 6 P.M. local, during standard and daylight savings time.

Geographic position of airport. The location of the airport is expressed in degrees, minutes, and tenths of minutes of latitude and longitude. For example: N37°43.28′W122°13.24′.

Charts. The sectional and enroute low- and high-altitude charts, and panel, on which the airport or facility can be found are indicated. Helicopter chart locations are shown as COPTER. In Figure 8-3, our airport can be found on the San Francisco sectional, panel A of the H-2 enroute high-altitude, panel F of the L-2 enroute low altitude chart, and the IFR area chart A.

Instrument approach procedures. Instrument approach procedure (IAP) indicates that a public-use, FAA instrument approach procedure has been published for the airport.

Airport sketch. The airport sketch, when provided, depicts the airport and related topographical information as seen from the air and should be used in conjunction with the text. Symbology is similar to the airport sketch on IAP charts. Additional aeronautical and topographical features are illustrated in Fig. 8-4. AD indicates that an airport diagram is available in the "Airport Diagrams" section of the A/FD.

Elevation. Elevation, given in feet above MSL, is the highest point on the landing surface, and is never abbreviated. When elevation is sea level it will be indicated as (00), below sea level a minus (−) will precede the figure. In Figure 8-3, the airport elevation is 6 ft MSL.

Rotating light beacon. The letter B indicates the availability of a rotating beacon. These beacons operate dusk to dawn unless otherwise indicated in airport remarks.

Servicing. Available services are represented by code:

- S1 Minor airframe repairs
- S2 Minor airframe and minor powerplant repairs

RUNWAYS	Light Plane, Ski or Water Landing Area	AIRPORT BEACON	
TREES		WIND CONE	
CUTS AND FILLS	Cut Fill	LANDING TEE	
CLIFFS AND DEPRESSIONS		TETRAHEDRON	
DITCH		CONTROL TOWER	
HILL		HELICOPTER LANDING AREA	H

8-4 *The airport sketch, when provided, depicts the airport and related topographical information as seen from the air and should be used in conjunction with the text.*

- S3 Major airframe and minor powerplant repairs
- S4 Major airframe and major powerplant repairs

Fuel. Availability and grade of fuel are also coded:

- 80 Grade 80 gasoline (red).
- 100 Grade 100 gasoline (green).
- 100LL Grade 100 low-level gasoline (blue).
- A Jet A kerosene, freeze point −40°C.
- A1 Jet A-1 kerosene, freeze point −50°C.
- A1+ Kerosene with icing inhibitor, freeze point −50°C.
- B Jet B wide-cut turbine fuel, freeze point −50°C.
- B+ Jet B wide-cut turbine fuel with icing inhibitor, freeze point −50°C.

- MOGAS Automobile gasoline used as an aircraft fuel. (Automobile gasoline may be used in specific aircraft engines that are FAA certified. MOGAS indicates automobile gasoline, but grade, type, and octane rating are not published. Due to a variety of factors, the fuel listed might not always be obtainable to transient pilots. Confirmation of availability should be made directly with fuel vendors at planned refueling locations.)

Oxygen. The availability of oxygen is indicated by one of the following:

- OX 1 High pressure
- OX 2 Low pressure
- OX 3 High pressure—replacement bottles
- OX 4 Low pressure—replacement bottles

Traffic pattern altitude. The first figure shown is traffic pattern altitude (TPA) above MSL, the second figure, in parentheses is TPA above airport elevation. In Figure 8-3, the pilot is referred to the remarks section for TPA.

Airport of entry and landing rights airport. Airport of entry (AOE) is a customs airport of entry where permission from U.S. Customs is not required, but at least 1 hour advance notice of arrival must be furnished. A landing rights airport (LRA) is one where application for permission to land must be submitted in advance to U.S. Customs, and at least 1 hour advance notice of arrival must be furnished. Advance notice of arrival at AOE and LRA airports may be included in the flight plan when filed in Canada or Mexico, where flight notification service (ADCUS) is available. Airport remarks will indicate this service. This notice will also be treated as an application for permission to land in the case of an LRA. Although advance notice of arrival may be relayed to Customs through Mexico, Canada, and U.S. communications facilities by flight plan, the aircraft operator is solely responsible for ensuring that customs receives the notification. In our example, Oakland is a landing rights airport, LRA, and, notice in remarks, ADCUS is available.

Certificated airport (FAR 139). Airports serving Department of Transportation–certified carriers and certified under FAR Part 139 are indicated by the ARFF index, which relates to the availability of crash, fire, and rescue equipment. Index definitions are listed in the directory and FAR 139, "Certification and Operations: Land Airports Serving Certain Air Carriers." When the ARFF index changes, because of temporary equipment failure or other reasons, a NOTAM D will be issued advertising the condition.

FAA inspection. All airports not inspected by the FAA will be identified by the note: Not insp. This indicates that airport information has been provided by the owner or operator of the field.

Runway data. Runway information is shown on two lines. Information common to the entire runway is shown on the first line while information concerning the runway ends is shown on the second or following line. Lengthy information will be placed in airport remarks. Runway directions, surface, length, width, weight-bearing capacity, lighting, gradient, and remarks are shown for each runway. Direction, length, width, lighting, and remarks are shown for seaplanes. The full dimensions of helipads are shown. Runway lengths prefixed by the letter H indicate that the runways are hard-surface concrete or asphalt. If the runway length is not prefixed, the surface is sod, clay, and the like. Runway surface composition is indicated in parentheses after runway length:

- AFSC Aggregate friction seal coat
- ASPH Asphalt
- CONC Concrete
- DIRT Dirt
- GRVD Grooved
- GRVL Gravel, or cinders
- PFC Porous friction courses
- RFSC Rubberized friction seal coat
- TURF Turf
- TRTD Treated
- WC Wire combed

Runway strength data are derived from available information and is a realistic estimate of capability at an average level of activity. It is not intended as a maximum allowable weight or as an operating limitation. Many airport pavements are capable of supporting limited operations with gross weights of 25 to 50 percent in excess of the published figures. Permissible operating weights, insofar as runway strengths are concerned, are a matter of agreement between the owner and user. When desiring to operate into any airport at weights in excess of those published, users should contact the airport management for permission. Runway weight bearing capacity is indicated by code:

- S Single-wheel-type landing gear (DC-3)
- D Dual-wheel-type landing gear (DC-6)
- T Twin-wheel-type landing gear (DC-6, DC9)
- SBTT Single-belly twin tandem landing gear (DC-10)
- DT Dual-tandem-type landing gear (B707)
- TT Twin-tandem-type landing gear (B-52)
- DDT Double dual-tandem-type landing gear (B747)
- TDT Twin delta-tandem landing gear (C-5, Concorde)
- AUW Maximum weight-bearing capacity for any aircraft irrespective of landing gear configuration
- SWL Single wheel loading
- PSI Pounds per square inch, maximum PSI runway will support

Quadricycle and dual-tandem are considered virtually equal for runway weight-bearing consideration, as are single-tandem and dual-wheel. The omission of weight-bearing capacity indicates information is unknown. Three zeros are added to the figures for gross weight capacity. For example, S-90 single-wheel-type landing gear, weight 90,000 pounds.

Lighting available by prior arrangement only or operating part of the night only, or pilot controlled, and with specific operating hours is indicated under airport remarks. Because obstructions are usually lighted, obstruction lighting is not included in the lighting code. Unlighted obstructions on or surrounding an airport will be noted in airport remarks. Runway light nonstandard (NSTD) are systems for which the light fixtures are not FAA approved—color, intensity, or spacing does not meet FAA standards. Nonstandard lighting will be shown in airport remarks. Types of lighting are shown with the runway or runway end they serve. Lighting contractions can be decoded by referring to Fig. 8-4 and App. A.

The type of visual approach slope indicator (VASI) and its location are described by a three-digit alphanumeric code. The code begins with the letter V, indicating a VASI system. The next figure indicates the number of boxes utilized (2 = two boxes; 4 = four boxes; 6 = six boxes). The last letter indicates which side of the runway has the unit when it is a single-side installation (L, left; R, right). For example, V6R would be a 6-box VASI on the right side of the runway; V16 is a 16-box VASI on both sides of the runway.

Runway gradient will be shown only when it is 0.3 percent or more. When available, the direction of upward slope will be indicated. Lighting systems, obstructions, and displaced thresholds will be shown on the specific runway end. Right-hand traffic patterns for specific runways are indicated by "Rgt tfc."

In Figure 8-3, RWY 11-29 is hard surface, 10,000 × 150 feet. The surface is asphalt with porous friction courses. Wheel-bearing capacity is single-wheel 200,000, dual-wheel 200,000, dual-tandem 300,000, and double dual-tandem 900,000 pounds. High-intensity runway lights and runway centerline lights are available. Runway 11 has MALSR approach lighting system, right traffic pattern. Runway 29 is equipped with ALSF2 approach and touchdown zone lights.

Airport remarks. Airport remarks provide supplemental information on data already shown, or additional airport information. Data are confined to operational items affecting the status and usability of the airport.

Weather data sources. This section indicates the availability of weather data or an automated weather observing system (AWOS). AWOS is available in one of four systems:

- *AWOS-A:* reports altimeter setting only.
- *AWOS-1:* reports altimeter setting, wind, and usually temperature, dewpoint, and density altitude.
- *AWOS-2:* data in AWOS-1, plus visibility.

- *AWOS-3:* data in AWOS-1, plus visibility and cloud/ceiling information.

Other types of weather information are denoted by one of the following contractions:

- SAWRS Supplemental aviation weather reporting station for current weather information.
- LAWRS Limited aviation weather reporting station for current weather information.
- LLWAS Low-level wind shear alert system.
- HIWAS Under radio aids to navigation—hazardous inflight weather advisory service, a continuous broadcast of SIGMETs and AIRMETs, and urgent pilot reports.
- SWSL Supplemental weather service location providing current local weather via radio and telephone.
- TDWR Indicates airports that have terminal doppler weather radars.

When the automated weather source is broadcast over an associated airport NAVAID, it will be indicated in bold type (**HIWAS** 116.8 OAK).

Communications. Communications are listed in the following order along with the frequency:

- CTAF Common traffic advisory frequency
- ATIS Automatic terminal information service
- UNICOM Aeronautical advisory station
- AUNICOM Automated aeronautical advisory station
- FSS Flight service station
- APP CON Approach control; R indicates the availability of radar
- Tower Control tower
- GND CON Ground control
- GCO Ground communications outlet
- DEP CON Departure control
- CLNC DEL Clearance delivery
- PRE TAXI CLNC Pretaxi clearance

Automated UNICOM is a computerized, command response system that provides automated weather, radio check capability, and airport advisory information selected from an automated menu by microphone clicks. This feature is not shown on NACO IAP charts. A ground communications outlet (GCO) is a remotely controlled communications facility. Pilots at uncontrolled airports may contact ATC and FSS via a radio/telephone link. Pilots will use "four key clicks" to contact the ATC facility or "six key clicks" to contact the FSS—similar to PCL. This system is intended to be used only on the ground. Pretaxi clearance procedures have been established at certain airports to allow pilots of departing IFR aircraft to receive the IFR clearance before taxiing for takeoff. The availability of radar is indicated by the circled letter R.

Airspace. This section indicates the type of airspace (Class B, C, D, or E) and who provides the service. In the example, Class C service is provided by "Bay approach". For airports served by part-time towers, this section will give the type of airspace, and base of that airspace, in effect when the tower is closed.

Radio aids to navigation. The directory lists all NAVAIDs, except military TACANs, that appear on NACO visual or IFR charts, and those for which the FAA has approved instrument approach procedures. NAVAIDs within the National Airspace System have an automatic monitoring and shutdown feature in the event of malfunction. Unmonitored (UNMON) means that an ATC facility cannot observe the malfunction or shutdown if the facility fails. NAVAID NOTAM files are listed on the radio aids to navigation line. At times this NOTAM file will be different from the airport NOTAM file. For example, for the Watsonville localizer and NDB the NOTAM file is WVI. The Salinas VOR NOTAM file is SNS.

Radio class designators and standard service volume (SSV) classifications, discussed in Chap. 5, are listed: (T) terminal, (L) low altitude, and (H) high altitude. In addition to SSVs, restrictions within the normal altitude or range of a NAVAID are published. For example, "VOR unusable 030°-090° beyond 30 nautical miles below 5,000' indicates that the VOR cannot be relied upon for navigation between the 030 and 090 radials beyond 30 nm, below an altitude of 5000 ft.

Latitude and longitude coordinates, relation of the NAVAID to the airport, facility elevation, and magnetic variation are provided. In Figure 8-3, the facility is at the field and has an elevation of 10 ft with a magnetic variation of 17°E (at fld.10/17E).

ASR/PAR indicates that surveillance (ASR) or precision (PAR) radar instrument approach minimums are published. The availability of HIWAS is also listed.

Pertinent remarks concerning communications and NAVAIDs are included in this section. For example, possible interference to approach aids due to aircraft taxiing in the vicinity of the antenna, nonavailability of the emergency frequency at the tower, or tower local control sectorization will be listed in this section.

Refer to Figure 8-3. Let's translate some of the airport data beginning with information for Runway 09R-27L. This runway is 6212 × 150 ft, hard surfaced, consisting of asphalt with porous friction courses. It has high-intensity runway lights. The runways are served with a visual approach slope indicator (VASI), a four-box VASI on the left side of the runway (V4L). The glide angle is 3° and crosses the 09R threshold at 46 ft and 27L threshold at 55 ft. Trees are close to the final approach of 09R.

From the airport remarks section we see that 09R-27L is closed to four-engine wide body aircraft except by prior approval (PPR). All turbojet/fan, four-engine, and turboprop aircraft with weight over 12,500 pounds are prohibited from takeoff 27L or landing 09R between 10 P.M. and 4 A.M. local time (0600–1400‡). Runway 27L has distance remaining signs on the left side. The traffic pattern altitude 27L is 606 ft MSL or 600 ft AGL.

Alaska supplement

The *Alaska supplement* (AK) provides an A/FD for the state of Alaska, a joint civil/military flight information publication, and also, a FLIP-A/FD for Alaskan civil and military visual and instrument charts. The *Alaska supplement* contains the following sections:

- General information
- A/FD legend
- A/FD
- Notices
- Associated data
- Procedures
- Emergency procedures
- Position reports

General information, legend, and directory contain generally the same information as the A/FD, except that they include military and private airports. The legend and directory also contain information on jet aircraft starting units, military specifications for aviation fuels and oils, military oxygen specifications, and arresting gear.

Facilities covered by the FAA and DOD NOTAM system are indicated by a section (§) symbol for FAA/DOD NOTAMs or the section symbol for civil NOTAMs only. Pilots flying to airports not covered by the NOTAM system should contact the nearest flight service station or the airport operator for applicable NOTAM information.

Airports in the supplement are classified into two categories: military/federal government and civil airports open to the general public, plus some selected private airports. Airports are identified by an abbreviation:

- A U.S. Army
- AF U.S. Air Force
- ANG U.S. Air National Guard
- AR U.S. Army Reserve
- CG U.S. Coast Guard
- DND Canadian Department of National Defense
- FAA Federal Aviation Administration
- MC U.S. Marine Corps
- MOT Canadian Ministry of Transport
- N U.S. Navy
- NG U.S. Army National Guard

- PVT Private use only, closed to the public
- NMFS National Marine Fisheries Service
- USFS U.S. Forest Service

If there is no classification, the airport is open to the general public.

Airport lighting is indicated by number:

1. Portable lights—electrical
2. Boundary lights
3. Runway floods
4. Runway or strip
5. Approach lights
6. High-intensity runway lights
7. High-intensity approach lights
8. Sequenced flashing lights (SFL)
9. Visual approach slope indicator system (VASI)
10. Runway end identifier lights (REIL)
11. Runway centerline lights (RCL)
12. Precision approach path indicator (PAPI)
13. Optical landing system (OLS)

An L by itself indicates temporary lighting such as flares, smudge pots, or lanterns. An asterisk preceding an element indicates that it operates on request only, by phone, telegram, radio, or letter. Otherwise, lights operate sunset to sunrise, except where pilot-controlled lighting (PCL) is indicated.

In addition to the information contained in the A/FD, the *Alaska supplement* provides unique data for the area it serves. In Alaska, some FAA flight service stations provide long-distance communications: air/ground and a weather broadcast. These are published in the supplement. The supplement also provides military air refueling and military training route data.

The procedures section of the supplement contains weather/NOTAM procedures, ARTCC communications, military and civilian air defense identification zone (ADIZ) information, and other general data. A separate section provides emergency procedures. This section contains air intercept signals, air/ground emergency signals, and search and rescue procedures. The back cover contains position report, flight plan, and change of flight plan sequences for in-flight operations.

Pacific chart supplement

The *Pacific chart supplement* (PAC) is a civil flight information publication. It serves as an A/FD for the state of Hawaii and those areas of the Pacific served by United States facilities (American Samoa, Kiribati–Christmas Island; Tern, Kure, and Wake Island; and the Caroline, Mariana, and Marshall Islands). The supplement contains ATC procedures for operating in the Pacific, including the same information found in domestic terminal procedures publications, for its area of coverage. The Pacific chart supplement contains the following:

- General information
- A/FD legend
- Airport/FD
- Notices
- Associated data
- Procedures
- Emergency procedures
- Airport sketches
- Terminal procedures
- Position reports

General information, legend, and directory are similar in content and format to the domestic directory and the *Alaskan supplement.* Notices are divided in special, general, and area categories. Special notices include information of a permanent or temporary nature, and sectional chart corrections. General notices include navigational warning areas, preferred routes, and general information on flying to Hawaii. Area notices provide general information for operations in the covered areas, including terminal area graphics and Hawaiian island reporting service. Associated data contain NAVAID and communications information, VOR receiver checkpoints, parachute-jumping areas, special use airspace, visual navigation chart bulletin, and military training routes.

Procedures provide information on oceanic navigation and communications requirements, oceanic position reports, routes to the U.S. mainland, and SCATANA (security control of air traffic and air navigation aids) and ADIZ (air defense identification zone) procedures. Emergency procedures and airport sketches contain the same information as the *Alaska supplement.* The final section of the supplement contains terminal procedures. This section contains the same information in the same format as the domestic terminal procedures publication. DPs, STARs, airport diagrams, and instrument approach procedures are contained in this section.

The Notice to Airmen (NOTAM) publication

Failure to check NOTAMs has led many a pilot into an embarrassing and potentially hazardous situation. Increased use of direct user access terminals and other commercially available briefing systems means that interpreting and understanding NOTAMs have taken on a greater significance. Recent changes in the FAA's NOTAM system with the introduction of ICAO contractions and date/time groups make NOTAMs even more important. These factors will challenge pilots to a much greater degree than in the past.

The Federal Aviation Administration advertises the status of components or hazards in the National Airspace System (NAS) through aeronautical charts, the *Airport/Facility Directory,* other publications, and the National Notice to Airmen System. Changes are normally published on charts, in the Directory, or appear in the *Notices to Airmen* publication. Published NOTAMs are sometimes referred to as Class II. Class II is the international term used to identify NOTAMs that appear in printed form for mail distribution. The need for current charts and publications cannot be overemphasized. Recall the pilot who called flight service for a briefing from Bishop to Santa Cruz, California, where the airport had been closed for 2 years!

Published every 28 days, the *Notice to Airmen* publication is shown in Fig. 8-5. This publication is now available on the Internet at www.faa.gov/ntap. New or revised items will be indicated by shaded text. FDC NOTAMs, other than temporary flight restrictions, are incorporated in the publication. Once placed in the publication, FDC NOTAMs are removed from the FAA's telecommunications system. NOTAMs of a permanent nature are carried only until published on the appropriate aeronautical chart or in the *Airport/Facility Directory*. Therefore, pilots must check this document for current information. The publication is divided into four parts.

Part 1 is divided into three sections. Section 1 contains airway NOTAMs, sorted alphabetically by ARTCC and descending FDC NOTAM numerical order. Section 2 provides airport,

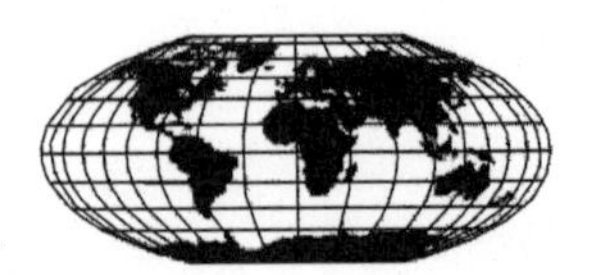

NOTICES TO AIRMEN

Domestic/International

Effective April 18, 2002 through May 15, 2002

Special Events In This Issue	Contents In Every Issue
Sporting Events	**General Information**
Montrose Ski Season	Foreword
Eagle Ski Season	Publication Schedule
Rifle Ski Season	NOTAM & Weather Contractions
Aspen Ski Season	Subscription Information
Telluride Ski Season	
Military Fly-By & Aerial Demonstrations Schedule	**Part 1.**
Thunder Over Louisville	Airway NOTAMs
Kentucky Derby	Airport, Facilities, & Procedural NOTAMs
Talladega 500 Winston Cup NASCAR	Content Criteria
The Winston & Coca-Cola 600 NASCAR	General FDC NOTAMs
Dover MBNA NASCAR	
Indianapolis 500	**Part 2.**
Jackson KMART 400 NASCAR	Part 95 Revisions to Minimum En Route IFR
Richmond Pontiac Excitement 400 NASCAR	Altitudes and Changeover Points
	Part 3. International
	General
	Flight Prohibitions, Potentially Hostile Situations, and Foreign Notices
	International Oceanic Airspace Notices
	Part 4.
	Graphic Notices

8-5 *The* Notices to Airmen *publication is available on the Internet.*

facility, and procedural NOTAMs. These include chart corrections, airport, facilities, procedural NOTAMs, and any other information required, listed alphabetically by state. Section 3 contains general FDC NOTAMs, those not listed under a specific LOCID. This information consists of flight advisories and restrictions. Excerpts from Part 1 are shown below.

FDC 9/1769 SMO FI/T SANTA MONICA, SANTA MONICA, CA.

NDB-B ORIG PROC NA.

This FDC was issued in 1999 (FDC *9*/1769). It was the 1769th FDC issued during that year (FDC 9/*1769*), and pertains to the Santa Monica airport (SMO). The information contained in the NOTAM is temporary (FI/T: flight information of a temporary nature). The SMO NDB-B original issuance approach procedure is not authorized (NA).

FDC 4/4047 STS FI/T SONOMA COUNTY SANTA ROSA, CA.

ILS RWY 32 AMDT 15...VOR RWY 32 AMDT 18...VOR/DME RWY 14 AMDT 1...CHANGE NOTE TO READ: WHEN CDSA NOT IN EFFECT, EXCEPT OPERATORS WITH APPROVED WEATHER REPORTING SERVICE, USE TRAVIS AFB /SUU/ ALTIMETER SETTING AND INCREASE ALL DH'S AND MDA'S BY 390 FEET.

This FDC changes a note on the three approaches listed to increase the decision heights and minimum descent altitudes by 390 ft when the Class D surface area (CDSA) is not in effect, unless the pilot has access to an approved weather reporting service. When Class D airspace is not in effect (the tower is closed) the altimeter setting is not available. Pilots without approved weather reporting service must use the Travis AFB (SUU) altimeter. The use of a remote altimeter setting requires an increase in minimums. (Automated weather observation systems solve this limitation.)

!FDC 9/0964 SJC FI/P SAN JOSE INTL SAN JOSE CA. CORRECT U.S. TERMINAL PROC VOL 2 OF 2 DATED 28 JAN 99 PAGE 401. GPS

RWY 30L ORIG...PLAN VIEW: HOLDING PATTERN AT SUNNE SHOULD
READ: 123 DEG INBOUND AND 303 DEG OUTBOUND.

This is an example of an error in the terminal procedures publications. It seems the FAA never has time to do it right, but always has time to do it over!

!FDC 8/1667 ZOA FI/T AIRWAY ZLA ZOA V25 SAN MARCUS /RZS/ VORTAC, CA TO POZOE INT, CA MEA 9500. POZOE INT, CA TO PASO ROBLES /PRB/ VORTAC, CA MEA 7000.

!FDC 4/2590 ZOA FI/T AIRWAY ZOA, NV. V165 MUSTANG (FMG) VORTAC, NV TO PYRAM INT, NV MOCA 10000.

These are examples of airway changes. Note that the LOCIDs are the appropriate ARTCC LOCIDs.

Part 2 contains revisions to 14 CFR Part 95 minimum enroute IFR altitudes and changeover points.

Part 3 incorporates international notices to airmen. This includes significant international information and data that may affect a pilot's decision to enter or use areas of foreign or international airspace. Foreign country data is listed alphabetically, followed by international oceanic airspace notices and United States overland/oceanic notices.

Part 4, graphic notices, contains special notices too long for other sections that concern a wide or unspecified geographical area, or items that do not meet other section criteria. Information in Part 4 varies widely, but is included because of its impact on flight safety.

Using the NOTAM system

Like METAR, PIREP, and TAF, NOTAM codes were changed in 1996. However, for NOTAMs the change was postponed several times. There were two major changes: the use of international contractions and a 10-digit date/time group.

The date/time group consists of year, month, day, and time (UTC). Times used on NOTAMs are now all UTC. The day begins at 0000Z and ends at 2359Z. For example, 9810291400 decodes as year 98 (1998), month 10 (October), day 29, and time 1400Z. A common contraction used with date/time groups is WEF. WEF translates as: "with effect from or effective from." Use care when determining effective times! These date/time groups can be very confusing. The absence of a date/time group means the condition is in effect and will continue until further notice (UFN). However, UFN is not transmitted in the NOTAM text. To indicate a condition is in effect and will exist until a specified time the contraction TIL (until), followed by a year/date/time group describes the effective period.

As part of the standard briefing, pilots receive any pertinent NOTAMs that are on hand. These would include NOTAMs on the status of NAVAIDs, airway changes, and airspace restrictions—NOTAM (D)s and FDC NOTAMs. If the FSS controller doesn't mention NOTAMs, ask. Briefers are human and NOTAMs are easy to overlook. NOTAMs are not necessarily provided during abbreviated or outlook briefings. Nor are NOTAMs normally available on the FSS's Telephone Information Briefing System (TIBS). Information contained in the *Notices to Airmen* publication is provided only on request.

A check of the *Airport/Facility Directory* should be a standard part of flight planning; significant information may be published. If we wish, request NOTAM (L)s from the tie-in FSS prior to descent and landing. FSS controllers are required to furnish any pertinent NOTAMs to landing aircraft within their flight plan area. Therefore, closing a flight plan with the tie-in FSS prior to landing will ensure receipt of any appropriate NOTAMs, as well as the altimeter setting, for uncontrolled airports. With the *Notice to Airmen* publication on the Internet, pilots have greater access to this document. However, if you don't have access to the *Notices to Airmen* publication, ask the briefer to check.

Pilots using DUAT or other commercially available systems may have to decode and translate NOTAMs. Even with systems that have a decode function, the pilot is still responsible for interpretation and application. Remember the DUAT disclaimer, "Non-associated FDC NOTAMs are available. Do you request them?" These would include temporary flight restrictions and airway changes. Commercial systems do not provide NOTAM (L)s, GPS RAIM NOTAMs, or information contained in the *Airport/Facility Directory* or the *Notice to Airmen* publication. The contents of these documents still remain the responsibility of the pilot. Should any doubt exist about the meaning or intent of a NOTAM, consult a flight service station for clarification.

The procedures discussed in this section apply equally to VFR and IFR flights. Only by understanding the system can pilots ensure they meet their regulatory obligation of obtaining all available information.

Checking NOTAMs is like going to the restroom before a flight. We know we should, but sometimes it's just not convenient. Both oversights can lead to a very uncomfortable flight!

By the way, earlier in the chapter, under "Directory legend," four location identifiers were presented. They decode as follows:

- VT11 Ass pirin acres
- 4OH4 Millertime
- OW15 Crash in international
- 84XS You asked for it, you got it (Toyota)

Some pilots think NOTAMs are a bunch of bull, sometimes they are. Believe it or not!

!CNO 09/004 CNO ARPT UNSAFE LOOSE BULL

9

Supplemental and international publications

Additional publications that support charts are published by the public and private sector-NIMA's flight information publications, *Canada flight supplement, Publicación de Información Aeronautica* for Mexico, and NACO supplemental documents. Private vendor publications are most often airport directories that offer supplemental airport information, such as an airport sketch and names, types and telephone numbers of airport, restaurant, lodging, and transportation services.

Flight information publications

The NIMA's equivalent to the *Airport/Facility Directory* are flight information publications (FLIPs), planning documents intended primarily for use in ground planning at base, squadron, and unit operations offices. They are revised between publication dates by issuing replacement pages or a planning change notice (PCN) on a schedule or as required basis: separate documents are general planning and area planning.

General planning contains general information. This publication is shown in Fig. 9-1. Below is a list of the information contained in this document.

- Index of aeronautical information
- Explanation of terms
- FLIP program
- Flight plans

- Pilot procedures,
- International civil aviation organization
- Aviation weather codes
- LORAN chart coverage
- Revisions/quality reports/requisitions/distribution/schedule requirements
- Operations and firings over the high seas

GP

DoD
FLIGHT INFORMATION PUBLICATION

GENERAL PLANNING

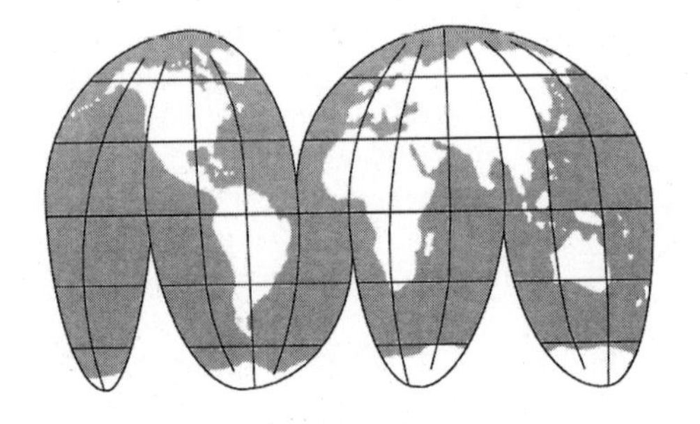

21 MARCH 2002

NEXT ISSUE 31 OCT 2002

PCN EFFECTIVE 11 JUL 2002

Consult NOTAMS for latest information.

Published by
NATIONAL IMAGERY AND MAPPING AGENCY
3200 SOUTH SECOND STREET
ST. LOUIS, MISSOURI 63118-3399

9-1 *The general planning document provides information that supports military planning, like the* Aeronautical Information Manual.

Area planning documents contain planning and procedural data for specific areas of the world. They include those theater, regional, and national procedures that differ from the standard procedures. Additionally, area planning military training routes are available.

As illustrated in Fig. 9-2, area planning documents are available for North and South America; Europe, Africa, and the Middle East; Pacific, Australasia, and the Antarctic; and eastern Europe and Asia. These publications supplement the visual, enroute, area, and terminal NIMA publications discussed in previous chapters.

Special-use airspace FLIPs are published in three books that contain tabulations of all prohibited, restricted, danger, warning, and alert areas. They also include intensive student jet training areas, military training areas, known parachute-jumping areas, and military operating areas. FLIP special-use airspace documents are available for the same areas of coverage as the area planning documents.

In addition to FLIP charts and publications, NIMA publishes a flight information handbook, plus the aeronautical chart updating manual (CHUM). The flight information handbook is a bound book containing aeronautical information required by DOD aircrews in flight, but is not subject to frequent change. Sections include information on emergency procedures, international flight data and procedures, meteorological information, conversion tables, standard time signals, and ICAO and NOTAM codes. The handbook is designed for worldwide use in conjunction with DOD FLIP enroute supplements.

NIMA publishes the CHUM semiannually with monthly supplements. The CHUM contains loran and miscellaneous notices and all known discrepancies to NIMA and most NACO charts affecting flight safety. Current chart edition numbers and dates for all NIMA charts are listed in the CHUM. The publication is intended for United States military use. NIMA also produces VFR and IFR enroute supplements. These serve the same purpose as the A/FD for the military.

Another volume, "Area Planning, Military Training Routes for North and South America, provides textual and graphic descriptions and operating instructions for all military training

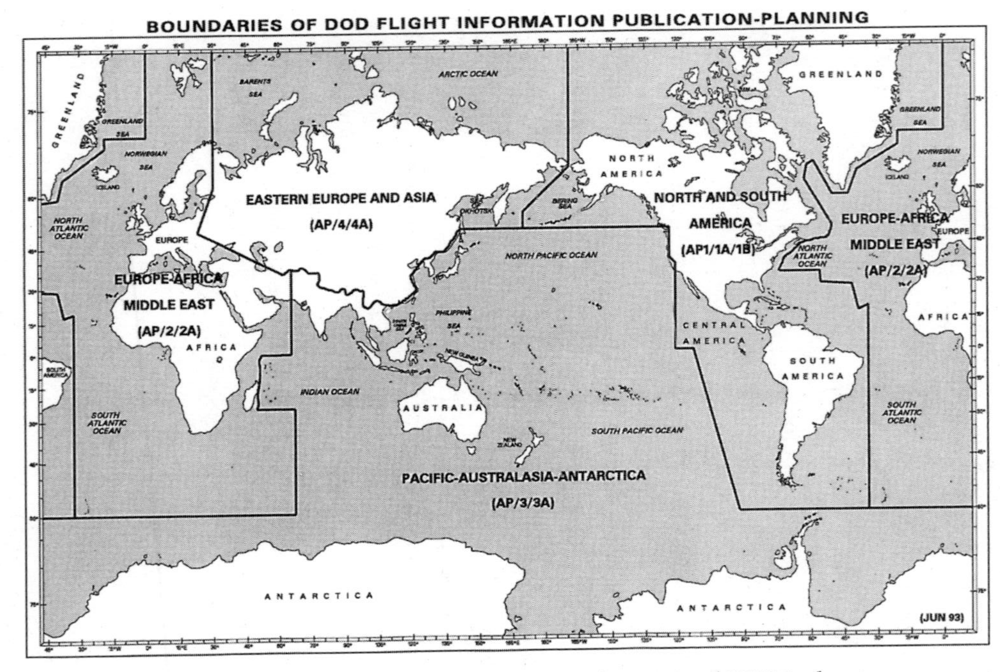

9-2 *FLIPs supplement visual, enroute, area, and terminal NIMA charts.*

routes, and refueling tracks. This publication supplements the area planning, military training routes chart discussed in Chap. 5. Each route and track is described by location, in radial distance from the nearest NAVAID, and altitude. Normal use times are also provided. This publication is available at any flight service station.

NIMA maintains a public sales program administered by NACO. Through NACO, NIMA has a free catalog that contains product descriptions, availability, prices, and order procedures for NIMA-produced aeronautical products. Although charts and publications are primarily of foreign areas, many domestic charts covering the United States are made available for purchase by the general public. NACO sales agents are located at or near principal civil airports worldwide. They may also be ordered directly by mail.

Use of any obsolete charts or publications for navigation is dangerous. Aeronautical information changes rapidly. It is critical that pilots have current charts and publications.

Aeronautical Information Manual (AIM)

The *Aeronautical Information Manual* contains basic information needed for safe flight in the United States National Airspace System. It includes chapters describing navigation aids, airspace, and the air traffic control system and provides information on flight safety and safe operating practices. It also includes a pilot-controller glossary. The AIM is designed to provide pilots with basic flight information and ATC procedures. As well as fundamentals required to fly in the system, it contains items of interest to pilots concerning health and medical factors.

The AIM is revised on a 112-day cycle, approximately three times a year. The complete manual is issued only once annually.

The AIM is the FAA's resource to help pilots understand operating in today's air traffic control system. In combination with most regulations, the AIM is the pilot's handbook to operating in the airspace. All of the navigational aids, lighting aids, and procedural descriptions, not within the scope of this book, are defined and explained in the AIM. I recommend that any serious pilot subscribe to, or purchase from one of the many private sources, a copy of this document. It is invaluable to the flight instructor, as well as the student pilot. The AIM is also available in an electronic version at www.faa.gov/atpubs. Features include:

- Navigation aids
- Aeronautical lighting and other airport visual aids
- Airspace
- Air traffic control
- Air traffic procedures
- Emergency procedures
- Safety of flight
- Medical facts for pilots
- Pilot controller glossary

Copies of the AIM can be obtained from the Government Printing Office (GPO). GPO will also provide, free of charge, subject bibliographies on a variety of aviation-related topics, which will guide the reader to government publications available through the Superintendent of Documents. Related subject bibliographies include: aircraft, airports, and airways; aviation information and training; and, weather. These bibliographies and a free catalog are available from:

Superintendent of Documents
U.S. Government Printing Office
Washington, D.C. 20402
Telephone: (202) 783-3238

Other supplemental products

Various additional supplemental products are available. Among these are a digital aeronautical chart supplement and a digital obstacle file.

The *Pilot's Handbook of Aeronautical Knowledge* contains essential information used in training. Subjects include the principles of flight, airplane performance, flight instruments, basic weather, navigation and charts, and excerpts from other flight information publications. This is a basic text, and ideal for the person getting started in aviation or interested in obtaining a general overview of aviation related topics.

The *Guide to Federal Aviation Administration Publications* is a 60-page document that contains information on the wide range of FAA documents and publications, and how they can be obtained. It lists available publications by category and gives the various sources. Listed also are civil-aviation-related publications issued by other federal agencies. Obtain a free copy by ordering FAA-APA-PG-9 from:

U.S. Department of Transportation
M-494.3
Washington, D.C. 20590

The *Location Identifiers Handbook,* no. 7350.5, lists the location identifiers authorized by the Federal Aviation Administration, Department of the Navy, and Transport Canada. It lists U.S. airspace fixes with latitude, longitude, rho-theta descriptions, and procedure codes. The handbook also includes guidelines for requesting identifiers and procedures for making assignments.

The *Contractions Handbook,* no. 7340.1, lists contractions for general aeronautical, National Weather Service, air traffic control, and aeronautical weather usage. The handbook provides encode and decode sections, plus air carrier, air taxi, and nationality identifiers: N (USA), C (Canadian), G (United Kingdom), and the like. The last sections provide a list of aircraft type contractions: C150, BE35, P28A, and the like.

Digital supplement

The digital aeronautical chart supplement (DACS) is designed to provide digital airspace data that are not otherwise readily available to the public. This publication was originally used only by air traffic controllers, but is now available to pilots for use in flight planning, and may soon be available on diskettes and tape. The following sections of the DACS are available separately or as a set:

- Section 1: High altitude airways—contiguous U.S.
- Section 2: Low altitude airways—contiguous U.S.
- Section 3: Reserved

- Section 4: Military training routes
- Section 5: Alaska, Hawaii, Puerto Rico, Bahamas, and selected oceanic routes
- Section 6: STARs and profile descent procedures
- Section 7: DPs
- Section 8: Preferred IFR routes
- Section 9: Air route and airport surveillance radar facilities

Features include the following:

1. Routes listed numerically by official designation
2. NAVAIDs and fixes listed by official location identifier
3. Fixes without official location identifiers (airway intersections, ARTCC boundary crossing points) listed by five digit FAA computer code
4. Latitude and longitude for each fix listed to tenths of seconds
5. Magnetic variations at NAVAIDs
6. Controlling ARTCC
7. Military training route descriptions (scheduling activity, altitude data, and route width)
8. Preferred IFR route (include departure or arrival airport name, and effective times)
9. Radar facilities (ground elevation, radar tower height and type of radar facility)
10. Data that is new or deleted since the last edition is clearly marked or listed

The NAVAID digital data file contains the geographic position, type, and unique identifier for every navigational aid in the United States, Puerto Rico, and the Virgin Islands. These data are chart independent and can be applied to a NACO chart for which the data are required. RNAV avionics can use these data without modification. The data is government certified and is compatible with the ARTCC system. This information is made available to the public, including avionics manufacturers, software developers, flight planning services, pilots, navigators, and other chart producers. Features include:

- NAVAID identifier
- Type NAVAID

- NAVAID status (commissioned or not commissioned)
- Latitude and longitude to tenths of a second
- Name of NAVAID
- NAVAID service volume category
- Frequency of NAVAID
- NAVAID elevation
- Magnetic variation
- ARTCC code where NAVAID is located
- State or country where NAVAID is located

These documents are used by chart producers and programmers of navigation systems. This information allows RNAV equipment to alert pilots of military and restricted areas and assist pilots in navigation.

Digital obstacle file

This quarterly file contains a complete listing of verified obstacles for the United States, Puerto Rico, and the Virgin Islands with limited coverage of the Pacific, Caribbean, Canada, and Mexico. Each obstacle is assigned a unique NACO numerical identifier. The obstacles are listed in ascending order of latitude within each state. A monthly revision file contains all changes made to verified obstacles during the previous 4-week period. The old record, as it appeared before the change, and the new record are shown. Features include:

- Unique NACO obstacle identifier
- Verification status
- State
- Associated city
- Latitude and longitude
- Obstacle type
- Number of obstacles
- Height AGL
- Height MSL
- Lighting
- Horizontal and vertical accuracy code

- Marking, if known
- FAA study number
- Julian date of last change

Commercial products

In addition to supplementary products from government sources, private vendors produce a variety of publications. These cover the entire spectrum from the copier quality reproductions of local ATC and UNICOM frequencies to high-quality airport sketch and data publications.

Jeppesen supplements its chart services with Federal Aviation Regulations and an airport and information directory, known as the JeppGuide. Features include:

- Radio aids to navigation, including coordinates, variation, and elevation
- Weather information sources
- Sunrise and twilight tables, and other common conversions
- A pilot controller glossary, airport and NAVAID lighting and markings, and services available to the pilot
- Federal Aviation Regulations
- International entry requirements
- Emergency procedures
- Airport diagrams and other airport information

A free catalog is available. Address and telephone numbers are contained in App. B.

The Aircraft Owners and Pilots Association (AOPA) publishes *Aviation USA,* available for sale to members and non-members, which provides a guide to AOPA member services, as well as an airport directory. It also has a section containing much of the material in the AIM. Publications of this type provide information on airport operators, transportation, lodging, and food services.

There are a number of commercially available publications similar to *Aviation USA,* but with subscription update service. These manuals provide a comprehensive listing of airports, with airport sketches; communication and NAVAID frequencies; and phone numbers for services, restaurants, and hotels and motels. The Oakland FSS subscribes to a publication that covers the state of California; telephone numbers for FBOs, hotels, and restaurants have been tremendous time savers in locating an overdue aircraft and a pilot that has inadvertently forgotten to close a flight plan.

Pilots should not overlook the value of these publications. Most airport pilot shops carry these products or they may be ordered from an aviation catalog.

International publications

Information for international flights is available from various public and private sources. Among these are the U.S. government's *International Flight Information Manual.* In addition to these documents, the United States Customs Service publishes the *Guide for Private Flyers* and *Customs Hints for Returning Residents.* Pamphlets are available for sale by the U.S. Government Printing office, and provide detailed customs requirements and procedures for pilots departing and returning to the United States.

The International Civil Aviation Organization (ICAO) provides extensive information for both private and commercial international aviation operations. Several private companies provide information, flight planning, meteorological, customs, and other international services to pilots.

International Flight Information Manual

The *International Flight Information Manual* is designed primarily as a preflight and planning guide for use by United States nonscheduled operators, businesses, and private aviators contemplating flights outside of the United States. This manual, which is complemented by the international NOTAM publication, contains foreign entry requirements; a directory of airports of entry, including data that are rarely amended; and pertinent regulations and restrictions. Information of a rapidly changing nature such as hours of operation, communications frequencies, and danger area boundaries, including restricted and prohibited areas, are not included and the pilot assumes the responsibility for securing that information from other sources: charts, NOTAMs, and enroute supplements. The basic manual is revised every April, with changes issued in July, October, and January. This manual is available at most flight service stations.

International NOTAMs, which are now a section in the *Notice to Airmen* publication, provides NOTAM service on a worldwide basis. It covers temporary hazardous conditions, changes in facility operational data, and foreign entry procedures and regulations. It supplements the international flight information manual, and is also available at most flight service stations. The respective international NOTAM publication and manual are available on a one year subscription service.

The *United States Aeronautical Information Publication* (AIP) is issued every 2 years with changes every 16 weeks. This is a comprehensive aeronautical publication containing regulations and data for safe operations in the National Airspace System. It is produced in accordance with the recommended standard of the International Civil Aviation Organization (ICAO).

Canada Flight Supplement

The *Canada Flight Supplement* is a joint civil/military publication issued every 56 days that contains information on airports, serving the same purpose as the United States' A/FD. It is published under the authority of NAV CANADA and the Chief of the Defense Staff, published by Geomatics Canada, Department of Natural Resources. The flight supplement contains the following sections:

- Special notices
- General section
- Aerodrome/facility directory
- Planning
- Radio navigation and communications
- Military flight data and procedures
- Emergency

Special notices direct the pilot's attention to new or revised procedures.

The general section contains information of a universal nature, such as procurement, abbreviations, and a cross-reference airport location identifier and name section. This section provides an airport and facilities legend for the supplement.

The aerodrome/facility directory serves the same purpose as its U.S. directory counterpart. This section might include an aerodrome sketch. The sketch depicts the airport and its immediate surrounding area. An obstacle clearance circle is provided to assist VFR pilots operating within close proximity to the airport, but should not be considered minimum descent altitudes. Altitudes shown are the highest obstacle, plus 1000 ft within a 5-nm radius. The last portion of this section contains a directory of North Atlantic airports and facilities (Azores, Bermuda, Greenland, and Iceland).

The planning section contains information on flight plan filing and position reporting, and supplementary data. It provides definitions of airspace classes, VFR chart updating data (aeronautical chart bulletin), and preferred IFR routes. A list of designated airway and oceanic control boundary intersection coordinates is provided.

Radio navigation and communications lists NAVAIDs by location and identifier. This section provides lists of marine radio beacons, commercial broadcasting stations, and other commonly used frequencies and services.

The military flight data and procedures section contains procedures and flight data for military operations in Canada and the North Atlantic.

The emergency section provides information similar to that contained in the Alaska supplement and Pacific chart supplement. Features include:

- Transponder operation
- Unlawful interference (hijacking)
- Traffic control light signals
- Fuel dumping
- Search and rescue

- Recommended procedures to assist in search
- Procedures when someone in distress is spotted
- Small craft distress signals
- Avoidance of search and rescue areas
- Emergency radar assistance
- Emergency communication procedures
- Two-way communications failure
- Information signals
- Military visual signals
- Interception of civil aircraft
- Interception signals
- Signals for use in the event of interception

Canada produces a water aerodrome supplement that provides tabulated and textual information to supplement Canadian VFR charts. This bound booklet contains an aerodrome/facilities directory of all water landing areas shown on Canadian VFR charts. Communications, radio aids, and associated data are also listed.

The *Canada Air Pilot* contains aeronautical information that is specifically pertinent to the conduct of the arrival or departure portion of flight, instrument approach procedures, standard instrument departure procedures, and noise abatement procedures. It is amended and reissued on a 56-day cycle. Seven volumes provide coverage across Canada, in a new throw-away format.

Pilots planning flights into Canada may obtain a pamphlet, *Air Tourist Information Canada,* free of charge. This pamphlet describes procedures for entering Canada and also lists pertinent aeronautical information publications. Pilots planning flights to Canada would be well advised to obtain the tourist pamphlet and chart and publications catalog. Publications are available on a one-time sale basis. Addresses, telephone numbers, and Internet sites are contained in App. B. Once you are in flight is no time to realize how helpful a particular chart or publication would be.

Publicación de Información Aeronautica

The Mexican government publishes the *Publicación de Información Aeronautica* (PIA)—in Spanish, of course. This document is similar to its United States and Canadian counterparts. Mexico publishes a series of VFR charts at the scale of 1:250,000, equivalent to the TAC series for the country. It also contains coverage of the ONC series at 1:1,000,000 scale. The PIA contains both low- and high-altitude enroute instrument charts.

The PIA contains an airport/facility directory. Features include:

- Abbreviations
- Time zones
- Sunrise-sunset tables
- Location identifiers
- Radio communications facilities
- Meteorological services
- Flight information region boundaries
- Flight information region frequencies
- Airspace information
- Search and rescue information
- Description of available charts
- Glossary of terms
- Aeronautical circulars

[A flight information region (FIR) is equivalent to a United States ARTCC.]

The PIA, like the U.S. *Pacific Supplement,* contains instrument approach charts. IAPs are similar in format to their United States, Canadian, and DOD equivalents, except in Spanish. The PIA also provides DPs, STARs, graphic depiction of terminal control areas, and airport diagrams. Addresses, telephone numbers, and Internet sites are contained in App. B of this book.

ICAO publications

The International Civil Aviation Organization publishes a catalog of publications and audiovisual training aids. Most

ICAO publications are available in English, French, Russian, and Spanish; Arabic is being introduced on a gradual basis. The catalog indicates which language versions are available. Below is a general description of ICAO publications.

- Conventions and related acts
- Agreements and arrangements
- ICAO rules of procedure and administrative regulations
- Annexes to the convention on international civil aviation
- Procedures of air navigation services
- Regional supplementary procedures
- Assembly
- Council
- Air navigation
- Air transport
- Legal
- Miscellaneous publications
- Indexes of ICAO publications
- Audiovisual training aids

ICAO publications cover all aspects of the organization from the convention and related acts to procedures for air navigation service, air navigation, and air transport rules. Many of the more pertinent documents are retained at FAA flight service stations for reference. A free catalog is available. addresses, telephone numbers, and Internet sites are contained in App. B.

Commercial international products

Various commercial vendors provide international flight planning services. Among them are the Aircraft Owners and Pilots Association (AOPA) and Jeppesen. AOPA provides international services to its members. Jeppesen provides its services per set fee.

AOPA has developed flight planning guides for Alaska, Canada, Mexico, and the Bahamas. These booklets include:

- Preflight planning and preparation
- Flying in the country

- Flight rules
- Returning to the United States
- Local information
- Airports of entry
- Supplemental information
- Special flight considerations

AOPA has information on training, services, and requirements for pilots planning to fly the North Atlantic. This includes addresses and telephone numbers for weather and flight planning vendors. Documents provided include various circulars, major changes in procedures and requirement, and the latest information available. AOPA is a resource that should not be overlooked. Addresses, telephone numbers, and Internet sites are contained in App. B.

In addition to aeronautical charts for the world, Jeppesen's international weather and flight planning services consist of:

- Jetplan
- Metplan
- NOTAM services
- International flight services
- OPSDATA service

Jetplan provides optimized flight plans that use the most fuel-efficient routes and altitudes. Their computerized database contains the world's low- and high-altitude routes, airports, DPs, and STARs. With worldwide winds and temperature and aircraft performance databases, the optimum routes and altitudes are obtained.

Metplan is Jeppesen's weather service. It provides all required weather products and weather information to meet flight operational needs. Jeppesen's weather information complies with FAA/FAR and ICAO requirements.

Jeppesen's NOTAM services provide pilot and flight operation personnel with regulatory flight information for the world. Selected information is designed to meet specific user needs.

International flight services provide the user with permits, ground handling, flight plans, weather, ground transportation, and accommodations for the flight.

Jeppesen's Opsdata service provides airport analysis and airport data services. This provides the user with maximum allowable weights and other related information for high-performance aircraft.

A

NOTAM/chart contractions

Contraction	Meaning	Agency
ABN	airport beacon	ICAO
ABV	above	ICAO
ACC	area control center (ARTCC)	ICAO
ACCUM	accumulate	FAA
ACFT	aircraft	ICAO
ACR	air carrier	FAA
ACT	active	ICAO
ADIZ	air defense identification zone	FAA
ADZD	advised	ICAO
AFD	airport facility directory	FAA
AGL	above ground level	ICAO
ALS	approach lighting system	ICAO
ALT	altitude	ICAO
ALTM	altimeter	FAA
ALTN	alternate	ICAO
ALTNLY	alternately	FAA
ALSTG	altimeter setting	FAA
AMDT	amendment	ICAO
AMGR	airport manager	FAA
AP	airport	ICAO
APCH	approach	ICAO
APLGT	airport lighting	ICAO
APP	approach control	ICAO
ARFF	aircraft rescue and fire fighting	FAA
ARR	arrive, arrival	ICAO

ASOS	automatic surface observing system	FAA
ASPH	asphalt	ICAO
ATC	air traffic control	ICAO
ATIS	automatic terminal information service	ICAO
AUTH	authority	ICAO
AUTOB	automatic weather reporting system	FAA
AVBL	available	ICAO
AWOS	automatic weather observing system	FAA
AWY	airway	ICAO
AZM	azimuth	ICAO
BA FAIR	braking action fair	ICAO
BA NIL	braking action nil	ICAO
BA POOR	braking action poor	ICAO
BC	back course	FAA
BCN	beacon	ICAO
BERM	snowbank/s containing earth/gravel	FAA
BLW	below	ICAO
BND	bound	FAA
BRG	bearing	ICAO
BYD	beyond	FAA
CAAS	Class A airspace	FAA
CAT	category	ICAO
CBAS	Class B airspace	FAA
CBSA	Class B surface area	FAA
CCAS	Class C airspace	FAA
CCLKWS	counterclockwise	FAA
CCSA	Class C surface area	FAA
CD	clearance delivery	FAA
CDAS	Class D airspace	FAA
CDSA	Class D surface area	FAA
CEAS	Class E airspace	FAA
CESA	Class E surface area	FAA
CFR	Code of Federal Regulations	FAA
CGAS	Class G airspace	FAA
CHG	change or modification	ICAO
CIG	ceiling	FAA
CK	check	ICAO

CL	centerline	ICAO
CLKWS	clockwise	FAA
CLR	clearance, clear(s), cleared to	ICAO
CLSD	closed	ICAO
CMB	climb	ICAO
CMSND	commissioned	FAA
CNL	cancel	ICAO
CNF	computer navigation fix	FAA
COM	communications	ICAO
CONC	concrete	ICAO
COP	changeover point	FAA
CPD	coupled	FAA
CRS	course	FAA
CTAF	common traffic advisory frequency	FAA
CTC	contact	ICAO
CTL	control	ICAO
DA	decision altitude	ICAO
DALGT	daylight	FAA
DCMSN	decommission	FAA
DCMSND	decommissioned	FAA
DCT	direct	ICAO
DEGS	degrees	ICAO
DEP	depart, departure	ICAO
DEPPROC	departure procedure	FAA
DH	decision height	ICAO
DISABLD	disabled	FAA
DIST	distance	ICAO
DLA	delay or delayed	ICAO
DLT	delete	FAA
DLY	daily	FAA
DME	distance measuring equipment	ICAO
DMSTN	demonstration	FAA
DP	dew point temperature	ICAO
DRFT	snowbank/s caused by wind action	FAA
DSPLCD	displaced	FAA
DUAT	direct user access terminal	FAA
E	east	ICAO
EB	eastbound	ICAO

EFAS	enroute flight advisory service	FAA
ELEV	elevation	ICAO
ENG	engine	ICAO
ENRT	enroute	ICAO
ENTR	entire	FAA
EXC	except	ICAO
FAC	facility or facilities	ICAO
FAF	final approach fix	ICAO
FAP	final approach point	FAA
FAN MKR	fan marker	ICAO
FBO	fixed-base operator	FAA
FDC	flight data center	FAA
FI/T	flight inspection temporary	FAA
FI/P	flight inspection permanent	FAA
FM	from	ICAO
FMS	flight management system	FAA
FNA	final approach	ICAO
FPM	feet per minute	ICAO
FREQ	frequency	ICAO
FRH	fly runway heading	FAA
FRZN	frozen	FAA
FSS	automated/flight service station	ICAO
FT	foot, feet	ICAO
GC	ground control	FAA
GCA	ground control approach	ICAO
GNSS/GLS	Global Navigation System/ Landing System	FAA
GCO	ground communication outlet	FAA
GOVT	government	FAA
GP	glide path	ICAO
GPS	Global Positioning System	FAA
GRVL	gravel	ICAO
HAA	height above airport	FAA
HAT	height above touchdown	FAA
HDG	heading	ICAO
HEL	helicopter	ICAO
HELI	heliport	FAA
HIRL	high-intensity runway lights	FAA
HIWAS	hazardous inflight weather advisory service	FAA
HLDG	holding	ICAO

HOL	holiday	ICAO
HR	hour	ICAO
IAF	initial approach fix	ICAO
IAP	instrument approach procedure	FAA
ICAO	International Civil Aviation Organization	ICAO
ID	identification	ICAO
IDENT	identify, identifier, identification	ICAO
IF	intermediate fix	ICAO
ILS	instrument landing system	ICAO
IM	inner marker	ICAO
IMC	instrument meteorological conditions	ICAO
IN	inch, inches	ICAO
INBD	inbound	ICAO
INDEFLY	indefinitely	FAA
INFO	information	ICAO
INOP	inoperative	ICAO
INSTR	instrument	FAA
INT	intersection	ICAO
INTL	international	ICAO
INTST	intensity	ICAO
IR	ice on runway/s	ICAO
KT	knots	ICAO
L	left	ICAO
LAA	local airport advisory	FAA
LAT	latitude	ICAO
LAWRS	limited aviation weather reporting station	FAA
LB	pound/s	FAA
LC	local control	FAA
LCTD	located	FAA
LDA	localizer-type directional aid	FAA
LDG	landing	ICAO
LGT	light or lighting	ICAO
LGTD	lighted	FAA
LIRL	low-intensity runway lights	FAA
LLWAS	low-level wind shear alert system	FAA
LLZ	localizer	ICAO

LM	compass locator at ILS middle marker	ICAO
LNAV	lateral navigation	FAA
LO	compass locator at ILS outer marker	ICAO
LOC	local, locally, location	ICAO
LOCID	location identifier	FAA
LONG	longitude	ICAO
loran	long-range radio navigation	FAA
LRN	long-range navigation	FAA
LSR	loose snow on runway/s	FAA
LT	left turn	FAA
MAA	maximum authorized altitude	FAA
MAG	magnetic	ICAO
MAINT	maintain, maintenance	ICAO
MALS	medium-intensity approach light system	FAA
MALSF	medium-intensity approach light system with sequenced flasher indicator lights	FAA
MALSR	medium-intensity approach light system with runway alignment	FAA
MAP	missed approach point	FAA
MAPT	missed approach point	ICAO
MCA	minimum crossing altitude	ICAO
MDA	minimum descent altitude	ICAO
MEA	minimum enroute altitude	ICAO
MEF	maximum elevation figure	FAA
MED	medium	FAA
MIN	minute(s)	ICAO
MIRL	medium-intensity runway lights	FAA
MLS	microwave landing system	ICAO
MM	middle marker	ICAO
MNM	minimum	ICAO
MOC	minimum obstruction clearance	ICAO
MOCA	minimum obstruction clearance altitude	FAA
MNT	monitor, monitoring, or monitored	ICAO

MRA	minimum reception altitude	ICAO
MSA	minimum safe altitude, minimum sector altitude	ICAO
MSAW	minimum safe altitude warning	FAA
MSG	message	FAA
MSL	mean sea level	ICAO
MU	mu meters	FAA
MUD	mud	FAA
MUNI	municipal	FAA
N	north	ICAO
NA	not authorized	FAA
NAV	navigation	ICAO
NAVAID	navigational aid	FAA
NB	northbound	ICAO
NDB	nondirectional radio beacon	ICAO
NE	northeast	ICAO
NGT	night	ICAO
NM	nautical mile/s	ICAO
NMR	nautical mile radius	FAA
NONSTD	nonstandard	FAA
NOPT	no procedure turn required	FAA
NOTAM	notice to airmen	FAA
NR	number	ICAO
NTAP	Notice to Airmen publication	FAA
NW	northwest	ICAO
OBSC	obscured, obscure, or obscuring	ICAO
OBST	obstruction, obstacle	ICAO
OM	outer marker	ICAO
OPR	operate, operator, or operative	ICAO
OPS	operation(s)	ICAO
ORIG	original	FAA
OROCA	off-route obstruction clearance altitudes	FAA
OTS	out of service	FAA
OVR	over	FAA
PA	precision approach	FAA
PAEW	personnel and equipment working	FAA
PAX	passenger/s	ICAO
PAPI	precision approach path indicator	ICAO

PAR	precision approach radar	ICAO
PARL	parallel	ICAO
PAT	pattern	FAA
PCL	pilot-controlled lighting	FAA
PERM	permanent	ICAO
PJE	parachute-jumping exercise	ICAO
PLA	practice low approach	ICAO
PLW	plow, plowed	FAA
PN	prior notice required	ICAO
PPR	prior permission required	ICAO
PRN	psuedo-random noise	FAA
PROC	procedure	ICAO
PROP	propeller	FAA
PSR	packed snow on runway/s	FAA
PTCHY	patchy	FAA
PTN	procedure turn	ICAO
PVT	private	FAA
RAIL	runway alignment indicator lights	FAA
RAMOS	Remote automatic meteorological observing system	FAA
RCAG	remote communication air/ground facility	FAA
RCL	runway centerline	ICAO
RCLL	runway centerline lights	ICAO
RCO	remote communication outlet	FAA
REC	receive or receiver	ICAO
RENL	runway end lights	ICAO
RELCTD	relocated	FAA
REP	report	ICAO
RLLS	runway lead-in light system	ICAO
RMNDR	remainder	FAA
RMK	remark/s	ICAO
RNAV	area navigation	ICAO
RNP	required navigational performance	ICAO
RPLC	replace	ICAO
RQRD	required	FAA
RRL	runway remaining lights	FAA
RSR	enroute surveillance radar	ICAO
RSVN	reservation	FAA

RT	right turn	FAA
RTE	route	ICAO
RTR	remote transmitter/receiver	FAA
RTS	return to service	ICAO
RUF	rough	FAA
RVR	runway visual range	ICAO
RVRM	runway visual range midpoint	FAA
RVRR	runway visual range rollout	FAA
RVRT	runway visual range touchdown	FAA
RWY	runway	ICAO
S	south	ICAO
SA	sand, sanded	ICAO
SAWRS	supplementary aviation weather reporting station	FAA
SB	southbound	ICAO
SCATANA	security control of air traffic and navigation aids	FAA
SDF	simplified directional facility	FAA
SE	southeast	ICAO
SFL	sequence flashing lights	FAA
SID	standard instrument departure	ICAO
SIMUL	simultaneous or simultaneously	ICAO
SIR	packed or compacted snow and ice on runway/s	FAA
SKED	scheduled or schedule	ICAO
SLR	slush on runway/s	FAA
SN	snow	ICAO
SNBNK	snowbank/s caused by plowing (windrow/s)	FAA
SNGL	single	FAA
SPD	speed	FAA
SSALF	simplified short approach lighting with sequence flashers	FAA
SSALR	simplified short approach lighting with runway alignment indicator lights	FAA
SSALS	simplified short approach lighting system	FAA
SSR	secondary surveillance radar	ICAO
SSV	standard service volume	FAA
STA	straight-in approach	ICAO

STAR	standard terminal arrival	ICAO
SUA	special-use airspace	FAA
SVC	service	ICAO
SVN	satellite vehicle number	FAA
SW	southwest	ICAO
SWB	scheduled weather broadcast	FAA
SWEPT	swept or broomed	FAA
T	temperature	ICAO
TAA	terminal arrival area	FAA
TACAN	tactical air navigational aid (azimuth and DME)	ICAO
TAR	terminal area surveillance radar	ICAO
TDZ	touchdown zone	ICAO
TDZE	touchdown zone	FAA
TDZLGT	touchdown zone lights	ICAO
TEMPO	temporary or temporarily	ICAO
TFC	traffic	ICAO
TFR	temporary flight restriction	FAA
TGL	touch and go landings	ICAO
THN	thin	FAA
THR	threshold	ICAO
THRU	through	ICAO
TIL	until	ICAO
TKOF	takeoff	ICAO
TRML	terminal	FAA
TRNG	training	FAA
TRSN	transition	FAA
TRSN	transient	FAA
TWR	airport control tower	ICAO
TWY	taxiway	ICAO
UNAVBL	unavailable	FAA
UNLGTD	unlighted	FAA
UNMKD	unmarked	FAA
UNMNT	unmonitored	FAA
UNREL	unreliable	ICAO
UNUSBL	unusable	FAA
UTC	coordinated universal time	FAA
VASI	visual approach slope indicator system	ICAO
VDP	visual descent point	FAA
VICE	in the place of	FAA
VIS	visibility	FAA

VMC	visual meteorological conditions	ICAO
VNAV	vertical navigation	FAA
VOL	volume	FAA
VOR	VHF omnidirectional radio range	ICAO
VORTAC	VOR and TACAN (collocated)	ICAO
W	west	ICAO
WAAS	wide-area augmentation system	FAA
WB	westbound	ICAO
WEF	with effect from or effective from	ICAO
WI	within	ICAO
WKDAYS	Monday through Friday	FAA
WKEND	Saturday and Sunday	FAA
WND	wind	FAA
WPT	waypoint	ICAO
WSR	wet snow on runway/s	FAA
WTR	water on runway/s	FAA
WX	weather	ICAO
Z	coordinated universal time	FAA
ZULU	coordinated universal time	FAA

B

Aeronautical charts and publications

This appendix contains a list of agencies that produce and distribute aeronautical charts and publications. Inclusion (or omission) from this list does not imply any recommendation (or lack of recommendation) of the product.

A Federal Aviation Administration/National Aeronautical Charting Office catalog is available from the FAA Distribution Division or selected authorized FAA aeronautical chart sales agents. The catalog describes available charting publications whose coverage goes beyond the geographic range of FAA products.

FAA/National Aeronautical Charting Office
Chart Sales Office
6501 Lafayette Avenue
Riverdale, Maryland 20737-1199
Tel: 800-638-8972
Tel: 301-436-8301
Internet: naco.faa.gov

Information on available Canadian charts and publications may be obtained from designated FAA chart agents or by contacting:

Canada Map Office
Department of Natural Resources
130 Bently Ave.
Nepean, Ontario
Canada K1A 0E9
Tel: 613-952-7000
Tel: 800-465-6277

Fax: 800-661-6277
Fax: 613-957-8861
Internet: ats.nrcan.gc.ca
E-mail: aero@nrcan.gc.ca

Information on available Mexican charts and publications may be obtained by contacting:

Dirección General de Aeronautica Civil (DGAC)
Departamento de Recursos Financieros
Calle de Nueva York No. 115
4to. Piso
Col. Napoles
C.P. 03810 Mexico, D.F.
Mexico
Tel: 52-5-682-1076 (682-1077)
Fax: 52-5-687-7325 (523-8310)

Dirección de Navigación Aereo
Blvd. Pureto Aereo 485
Zona Federal del Aeropuerto Internacional
15620 Mexico, D.F.
Mexico

Information on available Australian charts and publications may be obtained by contacting:

Australian Surveying and Land Information Group (AUSLIG)
Map Sales
P.O. Box 2
Belconnen ACT 2616
Australia
E-mail: mapsales@auslig.gov.au
Internet: www.auslig.gov.au

A free International Civil Aviation Organization (ICAO) catalog of publications and audiovisual training aids is available from:

International Civil Aviation Organization
ATTN: Document Sales Unit
999 University Street
Montréal, Quebec
H3C 5H7, Canada
Tel: 514-954-8022

Fax: 514 954-6769
E-mail: sales_unit@icao.org
Internet: www.icao.org/cgi/goto.plicao/en/sales.htm

The following are nongovernmental chart and publication providers.

Jeppesen
55 Inverness Drive East
Englewood, CO 80112-5498
Tel: 303-799-9090 or 800-621-5377
Fax: 303-328-4153
Internet: www.jeppesen.com

Howie Keefe's Air Charts System
12061 Jefferson Blvd., Suite A
Culver City, CA 90230-6219
Tel: 310-822-1996
Internet: www.airchart.com

Sporty's Pilot Shop
Clermont County/Sporty's Airport
Batavia, OH 45103-9747
Tel: 800-543-8633
Fax: 800-776-7897
Internet: sporty.com

The Aircraft Owners and Pilots Association (AOPA) provides chart service and has information on training, services, and requirements for pilots planning international flights.

Aircraft Owners and Pilots Association
Frederick Municipal Airport
421 Aviation Way
Frederick, MD 21701-4798
Tel: 301-695-2000 or 800-622-2672 (members)
Fax: 301-695-2375
Internet: www.aopa.org
E-mail: aopahq@aopa.org

Glossary

above ground level (AGL) Height, usually in feet, above the surface of the Earth.

above sea level (ASL) See mean sea level.

AGL See *above ground level.*

air defense identification zone (ADIZ) The area of airspace over land or water, within which the ready identification, location, and control of aircraft are required in the interest of national security.

aircraft approach category Landing minimums established for aircraft based on approach speed, and divided into A, B, C, D, E, and COPTER.

along track distance (ATD) Used on GPS approach charts to identify stepdown fixes that are not part of the database.

altimeter setting Corrects the altimeter for nonstandard pressure and temperature at field elevation. If the atmosphere above a station has a standard lapse rate (which it rarely does), indicated and true altitude above the station will be the same.

area navigation (RNAV) A method of navigation, either ground or satellite based, that permits aircraft operation on any desired course within the coverage of station-referenced navigation signals or within the limits of a self-contained system capability. Types include VORTAC, GPS, inertial, and loran.

areas Any area on the Earth's surface should be represented by the same area at the scale of the map.

ARFF See *certificated airport.*

Automated weather observing system (AWOS) A computerized system that measures some or all of the following: sky conditions, visibility, precipitation, temperature, dew point, wind, and altimeter setting.

Bernoulli effect The venturi effect of terrain that causes a decrease in air pressure, resulting in altimeter error.

bogs Areas of moist, soggy ground, usually over deposits of peat.

category See *aircraft approach category; ILS category.*

catenary A catenary, as depicted on aeronautical charts, is a cable, power line, cable car, or similar structure suspended between peaks, a peak and valley below, or across a canyon or pass.

certificated airport (14 CFR Part 139) An airport certified for commercial air carriers under Part 139, which relates to the requirements for crash, fire, and rescue equipment (ARFF).

changeover point (COP) The charted point where a pilot changes from one navigational facility to the next for course guidance.

chart A map that shows topographic and other information.

CHUM Chart Updating Manual, used to supplement National Imagery and Mapping Agency charts.

climb gradient A rate of climb in feet per nautical mile, usually associated with a requirement for a departure procedure.

common traffic advisory frequency (CTAF) A frequency designed for the purpose of airport advisory practices—pilot's position and intentions during takeoff and landing—at uncontrolled airports.

computer navigation fix (CNF) CNFs have been added to enroute, DP, STAR, and instrument approach procedures at mileage break points (turns) on airways, which formerly contained only the small letter x to indicate a mileage break for aircraft using database navigation system to identify airway turns. ATC will not request that aircraft hold at or report these fixes. Pilots should not request routes via CNFs, nor refer to CNFs in communications or flight plans.

Coordinated Universal Time (UTC) Formerly Greenwich Mean Time (GMT), also known as Z or ZULU time, UTC is the international time standard.

critical elevation The highest elevation in any group of related and more or less similar relief formations.

culture Features of the terrain that have been constructed by man, including roads, buildings, canals, and boundary lines.

datum Any quantity, or set of quantities, which may serve as a reference or base for other quantities.

decision altitude (DA) An ICAO term for a specified altitude in a precision approach, where a missed approach must be initiated if required.

decision height (DH) That altitude on a precision approach where the pilot must make the decision to continue the approach or execute a missed approach.

direct user access terminal (DUAT) A computer terminal where pilots can directly access meteorological and aeronautical information without the assistance of an FSS.

drainage patterns Drainage patterns are the overall appearance of features associated with water, such as shorelines, rivers, lakes, and marshes, or any similar feature.

DUAT See *direct user access terminal.*

feeder fix A fix depicted on instrument approach procedure charts that establishes the starting point of the feeder route.

feeder route A depicted course on instrument approach procedure charts to designate a route for aircraft to proceed from the enroute structure to an intermediate or initial approach fix.

final approach fix (FAF) The fix where the final approach segment begins.

final approach point (FAP) The point on an instrument approach where the final approach segment begins. This occurs when the final approach segment begins at the NAVAID used for final approach guidance, and no final approach fix is designated.

final approach segment That part of an instrument approach that begins at the final approach fix or point and ends at the missed approach point.

fixed base operator (FBO) A private vendor of airport services, such as fuel, repairs, and tiedown facilities.

flight level (FL) Altitude with the altimeter set to standard pressure (29.92 inches or 1013.2 millibars), pressure altitude. Flight level is the altitude reference for high-altitude flights, usually above 18,000 ft. (See also *QNE.*)

flight management system A computer system that uses a data base to allow routes to be preprogrammed and placed into the computer. The system is constantly updated with respect to position accuracy by reference to conventional navigation aids. The computer program automatically selects the most appropriate aids during the update cycle.

FLIP Flight information publication used to supplement National Imagery and Mapping Agency charts.

flumes Water channels used to carry water as a source of power to, for example, a water wheel.

Global Navigation System (GNS) landing system (GLS) A navigational system that provides precision navigation guidance for exact alignment and descent of aircraft on approach to a runway. GLS must have wide-area augmentation system (WAAS) equipment approved for precision approach (PA).

Global Positioning System (GPS) A navigational system based on low Earth-orbiting satellites.

great circle A circle on the surface of the Earth, the plane of which passes through the center of the Earth.

Greenwich meridian The meridian through Greenwich, England, serving as the reference for Greenwich time (now Coordinated Universal Time—UTC). It is accepted almost universally as the prime meridian, or the origin of measurements of longitude.

ground communication outlet (GCO) A remotely controlled communications facility. Pilots at uncontrolled airports may contact ATC and FSS via a radio/telephone link. Pilots will use "four key clicks" to contact the ATC facility or "six key clicks" to contact the FSS.

hachure A method of representing relief on a map or chart by shading in short disconnected lines drawn in the direction of the slopes. Hachures show the direction and degree of slope in hills and other elevations.

hazardous inflight weather advisory service (HIWAS) A continuous broadcast of hazardous weather conditions over selected NAVAIDs.

height above airport (HAA) The height of the minimum descent altitude above published airport elevation.

height above touchdown (HAT) The height of the decision height or minimum descent altitude above the highest runway elevation in the touchdown zone.

HIWAS See *hazardous inflight weather advisory service.*

horizontal datum A geodetic reference point that is the basis for horizontal control surveys, where latitude and longitude are known.

hummocks A wooded tract of land that rises above an adjacent marsh or swamp.

hydrography The science that deals with the measurements and description of the physical features of the oceans, seas, lakes, rivers, and their adjoining coastal areas, with particular reference to their use for navigational purposes.

hypsography The science or art of describing elevations of land surfaces with reference to a datum, usually sea level.

hypsometric tints A method of showing relief on maps and charts by coloring, in different shades, those parts that lie between selected levels.

ICAO See *International Civil Aviation Organization.*

ILS category The term *category,* with respect to the operation of aircraft, refers to a straight-in ILS approach. Category II and category III operations allow specially trained crews, operating specially equipped aircraft, using specially certificated ILS systems, lower landing minimums than are available with Category I.

initial approach fix (IAF) The point where an instrument approach begins.

initial approach segment That part of an instrument approach that begins at the initial approach fix and terminates at an intermediate fix.

intermediate segment That part of an instrument approach that begins at the intermediate fix and ends at the final approach fix.

International Civil Aviation Organization (ICAO) A specialized agency of the United Nations whose objective is to develop the principles and techniques of international air navigation and foster planning and development of international civil air transport.

isogonic lines Lines of equal magnetic declination for a given time, the difference between true and magnetic north.

Julian date The Julian calendar numbers the days of the year consecutively from 001, which is January 1; the year precedes the three-digit day group.

karst Topography feature that consists of sinkholes and fissures resulting from underground streams flowing through porous limestone, eroding the rock from below the ground.

Lambert conformal projection All meridians are straight lines that meet at a point beyond the map; parallels are concentric circles. Meridians and parallels intersect at right angles. A straight line from one point to another very closely approximates a great circle.

lateral navigation (LNAV) A navigational system that must have IFR approach approved WAAS, GPS, GPS-based FMS systems, or RNP-0.3 systems based on GPS or DME/DME.

Lateral navigation/vertical navigation (LNAV/VNAV) A navigational system that must have WAAS equipment approved for precision approach, or RNP-0.3 system based on GPS or DME/DME, with an IFR-approach-approved Baro-VNAV system.

latitude A linear or angular distance measured north or south of the Earth's equator.

location identifier (LOCID) Consisting of three to five alphanumeric characters, location identifiers are contractions used to identify geographical locations, navigational aids, and airway intersections.

longitude A linear or angular distance measured east or west from a reference meridian, usually the Greenwich meridian.

loran A long-range radio navigation position-fixing system using the time difference of reception of pulse-type transmissions from two or more fixed stations.

magenta A purplish red color used on aeronautical charts to distinguish different features.

mangrove Any of a number of evergreen shrubs and trees growing in marshy and coastal tropical areas; a nipa is a palm tree indigenous to these areas.

map A graphic representation of the physical features—natural, artificial, or both—of the Earth's surface, by means of signs and symbols, at an established scale, on a specified projection, and with a means of orientation. See also *chart*.

maximum authorized altitude (MAA) The highest altitude for which an MEA is designated, where adequate NAVAID signal coverage is assured.

maximum elevation figure (MEF) The MEF represents the highest elevation, including terrain or other vertical obstacles bounded by the ticked lines of the latitude/longitude grid on a chart.

mean sea level (MSL) Altitude above mean or average sea level. This is the reference altitude for most charted items. In Canada it is called *above sea level* (ASL).

Mercator projection The Mercator projection transfers the surface of the earth onto a cylinder tangent at the earth's equator. On the Mercator projection, meridians and parallels appear as lines crossing at right angles. (See also *transverse Mercator projection.*)

meridian A north-south reference line, particularly a great circle through the geographical poles of the Earth, from which longitudes are determined.

meridian graticule spacing Meridian graticule spacing, the network of parallels and meridians forming the map projection.

minimum crossing altitude (MCA) The minimum altitude at which a NAVAID or intersection may be crossed.

minimum descent altitude (MDA) The lowest authorized altitude on a nonprecision approach.

minimum enroute altitude (MEA) The minimum published altitude that assures acceptable navigational signal coverage, meets minimum obstruction clearance requirements, and ensures radio communications.

minimum obstruction clearance altitude (MOCA) The altitude that meets obstruction clearance criteria between fixes, but assures navigational signal coverage only within 22 nm of the NAVAID.

minimum reception altitude (MRA) The lowest altitude required to receive adequate NAVAID signals to determine specific fixes.

missed approach point (MAP) The point on an instrument approach where the missed approach procedure begins.

MSL See *mean sea level.*

NAVAID Any type of radio aid to navigation.

nipa A palm tree indigenous to marshy and coastal tropical mangrove. (See also *mangrove.*)

Nonperennial See *perennial.*

North American datum The horizontal datum for the United States developed in 1927, located at Meades Ranch, Kansas; referred to as the North American datum 1927 (NAD 27).

NOTAM file As used in a flight supplement, the associated weather or facility identifier (e.g., OAK) where notices to airmen for the associated facility will be located.

off-route obstruction clearance altitudes (OROCAs) An altitude, similar to a MEF, used on all U.S. low-altitude enroute charts to assist pilots using area navigation systems. OROCA is an off-route altitude that provides obstruction clearance with a 1000-ft buffer for nonmountainous terrain and a 2000-ft buffer for designated mountainous areas.

parallel A circle on the surface of the Earth, parallel to the plane of the equator and connecting all points of equal latitude.

penstock See *flumes.*

perennial A feature, such as a lake or stream, that contains water year round, as opposed to nonperennial, a feature that is intermittently dry.

planimetry The depiction of man-made and natural features, such as woods and water, that does not include relief.

polar stereographic projection A projection in which meridians are straight lines radiating from the pole and parallels are concentric circles. A rhumb line is curved and a great circle route is approximated by a straight line.

prime meridian See *Greenwich meridian.*

projection A systematic drawing of lines on a plane surface to represent the parallels of latitude and the meridians of longitude of the Earth.

QFE Altimeter setting by which the altimeter will read zero when the aircraft is on the ground, at field elevation.

QNE The altitude shown on the altimeter with the altimeter set to 29.92 inches, pressure altitude.

QNH Altitude above mean sea level displayed on the altimeter when the altimeter setting window is set to the local altimeter setting.

relief A representation of the inequalities of elevation and the configuration of land features on the surface of the Earth.

Required Navigational Performance (RNP) An ICAO standard for navigational systems based on GPS or DME/DME, with an IFR-approach-approved Baro-VNAV system.

rhumb line A line on the surface of the Earth cutting all meridians at the same angle.

runway visual range (RVR) The horizontal distance at which a pilot will be able to see high-intensity lights down the runway from the approach end.

Security Control of Air Traffic and Navigation Aids (SCATANA) A plan for the security control of civil and military air traffic and NAVAIDs under various emergency conditions.

scale The ratio of the distance on a chart to the corresponding distance on the surface of the Earth.

SCATANA See *Security Control of Air Traffic and Navigation Aids.*

scheduled weather broadcast (SWB) Available only in Alaska, a flight service station broadcast of certain meteorological and aeronautical information over NAVAID voice channels at 15 minutes past the hour.

special-use airspace (SUA) Airspace to which activities are confined because of their nature, such as military or national security operations. Restrictions to flight may be imposed, or pilots may be alerted to hazards of concentrated, high-speed, or acrobatic military flying.

spot elevation A point on a chart where elevation is noted, usually the highest point on a ridge or mountain range.

shaded relief A method of shading areas on a map or chart so that they would appear in shadow if illuminated from the northwest.

simplification The omission of detail that would clutter the map and prevent the pilot from obtaining needed information.

standard service volume (SSV) The distances and altitude for which a particular NAVAID can be relied upon for accurate navigational guidance.

terminal arrival area (TAA) Provides a transition from the enroute structure to the terminal environment for aircraft equipped with FMS or GPS. A TAA will not be found on all RNAV procedures. The TAA replaces the MSA for that approach procedure.

topography The configuration of the surface of the Earth, including its relief and the position of its streams, roads, cities, etc.

touchdown zone The first 3000 ft of the runway.

touchdown zone elevation (TDZE) The highest elevation in the first 3000 ft of the landing surface.

transverse Mercator projection A projection that rotates the cylinder so that it becomes tangent to a meridian. These projections are used in high north and south latitudes, and where the north-south direction is greater than the east-west direction. All properties of the regular Mercator projection are preserved, except the straight rhumb line.

tundra A rolling, treeless, often marshy plain, usually associated with arctic regions.

UNICOM A nongovernment communications facility that may provide airport advisory information.

UTC See *Coordinated Universal Time.*

vignette A gradual reduction in density so that a line appears to fade in one direction, often used to distinguish airspace boundaries.

visual descent point (VDP) The point on a nonprecision straight-in approach where normal descent from the minimum descent altitude to the runway touchdown point may begin.

wide-area augmentation system (WAAS) WAAS is a ground-based augmentation system for GPS positioning. It provides enhanced integrity and accuracy required for precision approaches.

ZULU (Z) See *Coordinated Universal Time.*

Index

About the Author

Terry T. Lankford is a retired FAA weather specialist, flight instructor, and holder of multiple pilot certifications. A resident of Pleasanton, CA, he is also the author of six McGraw-Hill aviation titles.